Technology, Knowledge and the Firm

Other Edward Elgar volumes in ASEAT Conference Proceedings Series:

Coombs, R., A. Richards, P-P. Saviotti and V. Walsh
*Technological Collaboration: The Dynamics of Cooperation in Industrial
    Innovation 1996*
ISBN: 1 85898 235 9

Coombs, R., K. Green, A. Richards and V. Walsh
Technological Change and Organization 1998
ISBN: 1 85898 589 7

Coombs, R., K. Green, A. Richards and V. Walsh
Technology and the Market: Demand, Users and Innovation 2001
ISBN: 1 84064 469 9

Technology, Knowledge and the Firm

Implications for Strategy and Industrial Change

Edited by

Ken Green

Professor of Environmental Innovation Management, Manchester Business School, University of Manchester, UK

Marcela Miozzo

Senior Lecturer in Innovation Studies, Manchester Business School, University of Manchester, UK

Paul Dewick

Lecturer in Technology Management, Manchester Business School, University of Manchester, UK

Edward Elgar

Cheltenham, UK • Northampton, MA, USA

Published by
Edward Elgar Publishing Limited
Glensanda House
Montpellier Parade
Cheltenham
Glos GL50 1UA
UK

Edward Elgar Publishing, Inc.
136 West Street
Suite 202
Northampton
Massachusetts 01060
USA

A catalogue record for this book
is available from the British Library

ISBN 1 84376 877 1

Printed and bound in Great Britain by MPG Books Ltd, Bodmin, Cornwall

Dedication

In memory of Albert Richards (organizer of the first five ASEAT conferences) for his friendship and support to the editors and of Keith Pavitt (of SPRU) for his intellectual leadership in the field of innovation studies.

Contents

List of contributors   ix
Acknowledgements   x

Introduction   1
Ken Green, Marcela Miozzo and Paul Dewick

Part One   Knowledge and the firm

1   Craft and code: intensification of innovation and management
    of knowledge   11
    Mark Dodgson, David M. Gann and Ammon Salter

2   The economics of governance: the role of localized knowledge
    in the interdependence among transaction, coordination and
    production   29
    Cristiano Antonelli

3   Innovation, consumption and knowledge: services and
    encapsulation   51
    Jeremy Howells

Part Two   Innovation and firm strategy

4   Paths to deepwater in the international upstream
    petroleum industry   73
    Virginia Acha and John Finch

5   Consumers and suppliers as co-producers of technology
    and innovation in electronically mediated banking:
    the cases of Internet banking in Nordbanken and
    Société Générale   92
    Staffan Hultén, Anna Nyberg and Lamia Chetioui

6   Technological shifts and industry reaction: shifts in fuel
     preference for the fuel cell vehicle in the automotive industry   126
     Robert van den Hoed and Philip J. Vergragt

7   Distant networking? The out-cluster strategies of new
     biotechnology firms   152
     Margarida Fontes

8   New science and old industries: adoption of biotechnology in
     European food companies   179
     Finn Valentin and Rasmus Lund Jensen

9   Commercialization of corporate science and the production
     of research articles   223
     Robert J. W. Tijssen

Part Three   Long-term technological change and the economy

10   Making (Kondratiev) waves: simulating long-run technical
     change for an Integrated Assessment system   253
     Jonathan A. Köhler

11   Nonlinear dynamism of innovation and knowledge transfer   267
     Masaaki Hirooka

Index   299

Contributors

Virginia Acha, University of Sussex, UK
Cristiano Antonelli, University of Turin, Italy
Lamia Chetioui, Ecole Centrale Paris, France
Paul Dewick, Manchester Business School, University of Manchester, UK
Mark Dodgson, University of Queensland, Brisbane, Australia
John Finch, University of Aberdeen, UK
Margarida Fontes, Instituto Nacional de Engenharia e Tecnologia Industrial, Portugal
David M. Gann, Imperial College London, UK
Ken Green, Manchester Business School, University of Manchester, UK
Masaaki Hirooka, Institute of Technoeconomics, Kyoto, Japan
Jeremy Howells, Centre for Research on Innovation and Competition (CRIC), University of Manchester, UK
Staffan Hultén, Stockholm School of Economics, Sweden
Rasmus Lund Jensen, Copenhagen Business School, Denmark
Jonathan A. Köhler, Tyndall Centre and University of Cambridge, UK
Anna Nyberg, Stockholm School of Economics, Sweden
Marcela Miozzo, Manchester Business School, University of Manchester, UK
Ammon Salter, Imperial College London, UK
Robert J. W. Tijssen, Centre for Science and Technology Studies (CWTS), Leiden University, the Netherlands
Finn Valentin, Copenhagen Business School, Denmark
Robert van den Hoed, Faculty of Industrial Design Engineering, Delft University of Technology, the Netherlands
Philip J. Vergragt, Tellus Institute, USA

Acknowledgements

Chapter 8 has been published in similar form as Valentin, F. and R. Lund Jensen (2003), 'Discontinuities and distributed innovation – The case of biotechnology in food processing', *Industry and Innovation* **10** (3), pp. 275–310. We gratefully acknowledge permission from Taylor and Francis (www.tandf.co.uk) to reproduce the above material.

Chapter 9 has been published in similar form as Tijssen, R. J. W., 'Commercialization of Corporate Science and the Production of Research Articles', *Research Policy*, forthcoming. We gratefully acknowledge permission from Elsevier to reproduce the above material.

The authors would like to thank the participants of the ASEAT conference, Manchester, April 2003. The authors would also like to thank Nirit Shimron for preparing the text and having the patience to deal with all the last-minute changes. Finally, we would like to thank Dymphna Evans of Edward Elgar for her support.

Introduction
Ken Green, Marcela Miozzo and Paul Dewick

This collection of essays brings together papers that were presented at the sixth biennial conference of Advances in Social and Economic Aspects of Technology (ASEAT) on 'Knowledge and Economic and Social Change: New Challenges to Innovation Studies' that was held in Manchester between 7 and 9 April 2003. The contributions have a common theme: the role of knowledge and innovation in firm strategy and industrial change. Underlying all the papers is an understanding that firms have distinctive ways of doing things and, moreover, that these ways of doing things have strong elements of continuity. The papers explore the role played by firms in developing, linking and utilizing knowledge produced in many social institutions to advance their organizational and technological capabilities. Understanding how firms advance their capabilities is essential to understanding how the economy operates and changes.

There is a long tradition of research underlining the importance of differences in organizational and technological capabilities of firms and their effect on economic performance. Edith Penrose's writings are the first point of departure to the understanding of how firms grow in the direction of their capabilities and how these capabilities expand and alter. Penrose (1959) saw the growth of a firm as based on the possession and development of unique and idiosyncratic resources. The second point of departure is George Richardson (1972) who presents firms as sets of activities which require knowledge, experience and skill in their performance.

Offering a different perspective to dominant equilibrium models and the competitive forces framework, other scholars have linked this broad capabilities perspective to questions of competition and industrial change. The importance of routines has been emphasized by evolutionary economics. Nelson and Winter (1977) have argued that R&D strategies of firms do not follow maximizing criteria but, instead, follow rules of thumb. They suggest that there are powerful intra-project heuristics when technology is advanced in certain directions, with payoffs from advancing in that same direction. The importance of routines has been further refined by the contributions on strategic management (Teece et al., 1997) that argue that the competitive advantage of firms is seen as resting on distinctive processes, shaped by the firm's assets and the evolution path it has adopted or inherited. These theories emphasize the significance of knowledge, and place the practices that surround the development and use of knowledge as fundamental elements defining the firm. The competitive advantage of firms is regarded as resting on distinctive high performance routines operating inside the firm, embedded in the firm's processes and conditioned by its history.

These contributions have helped advance a so-called 'dynamic capabilities' framework to examine the sources and methods of competitive advantage of firms operating in Schumpeterian environments of innovation-based competition. The work of business historians such as Chandler (1977) and Lazonick (1991) has added to this framework, contributing to the understanding of the way in which firms' organizational and technological capabilities have changed over the past century and how these changes explain the shifts in international industrial leadership.

Drawing on the above contributions, the chapters of this book take up the challenge of identifying the significance of mechanisms, internal or external to the firm, which can assist the development and use of resources, routines and capabilities through the management of knowledge. These give rise to patterns of technical change and knowledge bases underlying innovative activities, 'technological trajectories' (Dosi, 1988) and 'technological paradigms' (Freeman and Perez, 1988). More generally, they link to Schumpeter's (1934, 1942) work, which suggested that there are 'technological revolutions', which can have pervasive effects throughout the economy, leading not only to the emergence of new products, services, systems and industries but also affecting almost every other branch of the economy. These 'technological revolutions' affect the input cost structure and conditions of production and distribution, and the general 'natural trajectories' (Nelson and Winter, 1982) which, once established, exert a dominant influence on engineers, designers and managers. Schumpeter's long waves or cycles ('gales of creative destruction') can be seen as a succession of 'techno-economic paradigms' (Freeman and Perez, 1988) whose diffusion is accompanied by structural crises, in which social and institutional changes are required to bring about a better 'match' between the new technology and the institutions in the economy.

SYNOPSIS OF THE BOOK

Part One examines the mechanisms internal to the firm which contribute to the development of firm capabilities through the management of knowledge. Mark Dodgson, David Gann and Ammon Salter argue that the recent application of new electronic technologies in simulation and virtual modelling techniques in design and prototyping activities has resulted in the intensification of the innovation process. The authors ground their analyses on resource-based and behavioural theories of the firm, and strategic management theories on dynamic capabilities and evolutionary economics – all of which emphasize the significance of knowledge, and place the practices that surround the use of knowledge as fundamental elements of firm constructs. Also, they base their discussion on empirical research with engineering firms. They state that analyses in the innovation literature on the potential effect of these technologies have been limited. Technologies associated with design, particularly those built on ICT, have major implications for the cost and speed of the innovation process. Also, higher levels of involvement and integration of users and consumers facilitate industrial specialization and disaggregation. However, after considering the application of a number of ICT tools, they conclude that although ICT is important in its ability to codify some actions and behaviour, providing new mechanisms for the management of knowledge and innovation and development of routines and capabilities, it is the remaining tacit elements that define a firm's competitive advantage.

Cristiano Antonelli argues that the analysis of the role of knowledge in the economics of governance provides a framework that can integrate the research programmes of the economics of transaction costs and of the resource-based theory of the firm and overcome their limitations. He compares and contrasts the two theories and discusses the weaknesses of the approaches. He argues that transaction costs economics pays little attention to organizational knowledge and the competence necessary to coordinate and use the markets respectively and hence little room is left for understanding the process of accumulation of new organizational knowledge and the introduction of organizational innovation. The resource-based theory is unable to appreciate the role of organizational constraints in shaping the rate and direction of the growth of the firm. In the context provided by the economics of governance, he models the interdependence among transaction, coordination and production as a microsystem where localized technological and organizational knowledge plays a central role. The dynamic model presents the firm as a bundle of activities characterized by learning.

Jeremy Howells explores the role of consumption in the firm's innovation process, both in the existing literature and using primary case study material. The author argues that services play a key intermediary and conduit role in the innovation process and the analysis focuses on the role of services in the consumption of new goods by firms. The chapter briefly explores the trend by which firms are offering service products related to the manufactured goods they produce and takes examples from the vehicle manufacturer, aerospace and healthcare industries. These encapsulation mechanisms suggest a new concept of the relationship between consumption and innovation moving beyond the ideas of outsourcing and vertical integration. The chapter assesses the effect of the encapsulation process on innovation and suggests that firms need a more integrated consumption knowledge framework within which they can harness core capabilities.

Part Two examines the role of innovation and knowledge in the development of distinctive firm capabilities and strategy. Virginia Acha and John Finch draw upon the capabilities approach to examine recent changes in the upstream petroleum industry, in which a subset of operating firms have begun exploration, and, in some cases, production activities offshore in what the industry calls deepwater. Deepwater raises significant technical challenges, but also represents significant opportunities of large hydrocarbon accumulations. Such opportunities are widely believed to have been exhausted in shallow water offshore. The authors examine the tenacity of some large oil companies in maintaining these capabilities, instead of falling back on increasingly routine activities in mature fields, compared with the reluctance of other large firms to commit to deepwater activities. Drawing from case study evidence of three major firms involved in deepwater exploration and production, they identify counter tendencies to those of routinization and modularization. Building on Penrose, they argue that managers within these firms harness resources freed up by routinization and direct these to new non-routine activities.

Staffan Hultén, Anna Nyberg and Lamia Chetioui examine the different transitions from computer and telephone banking to internet banking in two retail banks, Nordbanken in Sweden (partner in the Nordic bank Nordea) and Société Générale in France; both of whom were well known for their successful use of computer banking. Their chapter focuses on the banks' technology choices and management of external resources and also assesses the role of customers and suppliers. The two case studies, informed by a series of interviews with managers in the two banks, identify different strategies in the choice of external partners and adoption rates based on path dependency, which ultimately determined the relative success of the banks' initiatives. The more immediate success of the banks' strategies is gauged by how many Internet banking customers they have managed to attract but the authors also consider the long term effects of the different trajectories Nordea and Société Générale are following.

Robert van den Hoed and Philip Vergragt examine the automotive industry's search for an alternative to the internal combustion engine. Increasingly the fuel cell vehicle (FCV) is seen as the sustainable alternative to the internal combustion engine and during the 1990s automotive programs on FC technology grew spectacularly. However, at present, there is no dominant design for the FCV, and the industry is split between using hydrogen, methanol or gasoline as the fuel for fuel cell vehicles. Using institutional theory and technology dynamics, the authors examine the environmental, infrastructural and technical consequences that underpin the industry's R&D decisions with regard to fuel preference. The authors draw on both qualitative and quantitative evidence from extensive interviews conducted with senior managers in the major car manufacturers to explore the factors behind the adoption decision. Institutional, strategic and cultural reasons are highlighted as important determinants in the technology choice decision and the authors identify a few opinion leaders who are shaping the decisions of the industry.

Margarida Fontes examines the characteristics of the biotechnology industry, focusing on the network structure of interfirm relationships that acts as coordination device between a variety of actors. While clustering is important for the evolution of this sector, biotechnology also presents some features – namely the international nature of scientific production and markets – that may facilitate firm development outside them. Through in-depth interviews with six new biotechnology firms in Portugal, the author identifies and discusses the main features of an 'out-cluster' strategy. She finds that these firms have been able to devise strategies to overcome some of the relative disadvantages of their location, enabling them to access and integrate nonlocal networks, to draw creatively from a combination of local and distant relationships and to manage this specific form of knowledge acquisition and business development.

Finn Valentin and Lund Jensen examine the distributed forms of innovation induced by biotechnologies in the food processing industry. The authors focus on the role of large firms and outside partners: universities, government research institutes and small specialist firms (Dedicated Biotechnology Firms, DBFs). In contrast to the US model of biotech success where DBFs have played a key role, industry incumbents have introduced virtually all innovations in the field of Lactic Acid Bacteria (LAB). Public research organizations (universities and government research institutes) are shown to contribute significantly to distributed R&D. Using a novel patent data mining tool, the authors build a complete map of contributions and collaborations from and between different institutions to describe the way in which a scientific discontinuity (in this case, biotechnology) shapes the emergence of a distributed organization of innovation and its subsequent evolution. Large incumbents are shown to patent at a level tenfold higher than general food processing firms; universities and, in particular, government research institutes are important collaborators in R&D. The findings of the chapter suggest that the US model of scientist–entrepreneurs–venture capitalists does not thrive in the area of LAB food biotechnology. The authors attribute this finding to low decomposability of the problem definition, which instead relies on an active role of public science.

In light of the downsizing or closing of research labs in many R&D-intensive firms over recent years, Robert Tijssen examines questions whether corporate research capabilities and activities are being managed increasingly as an economic asset, ruled by market forces. Despite a lack of global comparative measurements of industry's basic research efforts and its effect on research outputs, the author uses a new source of information on corporate research activity: research articles published in international scientific and technical journals. Important changes are identified from a statistical analysis of 290 000 corporate research articles that list author affiliate addresses in the corporate sector and which were published in international journals during the years 1996–2001. Whilst the number of patents and patent citations to research literature has increased significantly, the numbers of corporate research articles have declined steadily. The author undertakes a detailed analysis of trends in the pharmaceuticals sector and semiconductors sector and highlights sector-specific publication trends and patterns related to their innovation processes. Robert Tijssen concludes that the observed declines provide suggestive empirical evidence that corporate research is in a process of structural change where appropriation and commercialisation of research results reduce accidental and voluntary knowledge spillovers from industry into the public domain.

Part Three explores the relation between the internal mechanisms of the firm designed to develop and use knowledge and the long term patterns of technological change. Jonathan Köhler develops a simulation model of long term technical change. He argues that, due to deficiencies in data, the unsuitability of econometrics for modelling beyond the short to medium term as well as the number of socioeconomic variables to be considered, means that there is no generally accepted theory to date on long term technical change for incorporation into a macro-modelling structure. Based on Freeman and Louçã's (2001) descriptive theory, which encompasses the ideas on long waves from Kondratiev and Schumpeter, Kohler argues that socioeconomic activity since the late 1700s can be interpreted by a dynamic macroeconomic model. Learning by doing and falling production costs are combined with an investment bubble and a lagged supply response to generate the boom phase of a Kondratiev wave.

Building on the literature on diffusion of innovation, Masaaki Hirooka examines the period of technology development in firms, which occurs before innovation diffusion takes place (a phase which is more extensively researched than the technology development period). He argues that this period can be characterized by two trajectories: the technology trajectory and the development trajectory. The technology trajectory is composed of a series of core technologies (encompassing basic research) and the development trajectory is a locus of new products in the course of technology development (encompassing technology transfer from universities to research). Whilst it is well established that the diffusion of innovation has a nonlinear nature, Hirooka shows that the technology and development period can also be characterized by a logistic curve. Thus, the innovation paradigm is composed of three logistic trajectories describing the technology, development and diffusion stages. He offers evidence for all three trajectories and for the structure of the innovation paradigm. Considering the electronics paradigm, he identifies and discusses the key actors involved in the development trajectory: universities, venture businesses and the government.

REFERENCES

Chandler, A. (1977), *The Visible Hand: The Managerial Revolution in American Business*, Cambridge, MA: Harvard University Press.

Dosi, G. (1988), 'Sources, procedures, and microeconomic effects of innovation', *Journal of Economic Literature*, **26**, 1120–71.

Freeman, C. and F. Louçã (2001), *As Time Goes By*, Oxford: Oxford University Press.

Freeman, C. and C. Perez (1988), 'Structural crises of adjustment, business cycles and investment behaviour', in G. Dosi et al. *Technical Change and Economic Theory*, London: Pinter.

Lazonick, W. (1991), *Business Organization and the Myth of the Market Economy*, Cambridge: Cambridge University Press.

Nelson, R. R. and S. G. Winter (1977), 'In search of a useful theory of innovation', *Research Policy*, **6**, 36–76.

Nelson, R. R. and S. G. Winter (1982), *An Evolutionary Theory of Economic Change*, Cambridge, MA: Harvard University Press.

Penrose, E. (1959), *The Theory of the Growth of the Firm*, Oxford: Oxford University Press.

Richardson, G. B. (1972), 'The organization of industry', *Economic Journal*, **82** (327), 883–96.

Schumpeter, J. A. (1934), *The Theory of Economic Development*, Cambridge, MA: Harvard University Press.

Schumpeter, J. A. (1942), *Capitalism, Socialism and Democracy*, New York: McGraw-Hill.

Teece, D. J., G. Pisano and A. Shuen (1997), 'Dynamic capabilities and strategic management', *Strategic Management Journal*, **18** (7), 509–33.

PART ONE

Knowledge and the firm

1. Craft and code: intensification of innovation and management of knowledge

Mark Dodgson, David M. Gann and Ammon Salter

1. INTRODUCTION

The recent application of a range of new technologies used in simulation and virtual modelling techniques in design and prototyping activities has had significant implications for the innovation process (Dodgson et al., 2002; Schrage, 2000; Thomke, 2001). The use of these new tools, and the means by which these are integrated with other productive technologies in manufacturing and operations, have resulted in what we call the intensification of the innovation process, producing economies of effort and greater definiteness of aim in innovation.

In order to solve their design problems, engineers draw upon a vast body of knowledge about how things work (Vincenti, 1990). Even seemingly simple design requirements often have complex intellectual implications drawing upon routine knowledge used in 'normal' design coupled with the largely unknown experimentation carried out in 'radical' design activities. In this chapter we explore the implications of the new electronic toolkit for innovation, based upon the ways in which designers and engineers work in design, development, testing, production and coordination. We argue that detailed design relies upon tools that automate routines in stable areas of engineering. But we also contend that added value in design and development processes comes from creative and schematic work where designers produce radical solutions that are beyond the calculations embedded in routinized software programmes. They do this through 'conversations' and the use of 'visual cues', interacting with one another around new virtual product and process models. We suggest that this schematic work is about seeing the interfaces in the design process and obtaining commitment from different specialists. It is also about new forms of systems integration, effectively designing the process as well as the product (a subject which we shall address else-where).

Hitherto, much of the discussion about the use of Information and Communications Technologies (ICTs) in innovation processes has focused on the codification of knowledge. Although there is increasing realization of the potential impact of these technologies within the innovation litera-ture (Antonelli and Geuna, 2000; Pavitt, 2002; Steinmueller, 2000), much of the analysis has been limited. Some of the analysis has been deterministic and technology-led, introspective and uninformative, and almost all of it has been lacking in empirical evidence. Meaningful assessment of their impact and potential, we believe, depends upon an in-depth understanding of the nature and importance of prototyping and the design process in knowledge management and innovation. We shall argue here that the impli-cations of the technological changes we are experiencing will not be under-stood without a comprehensive understanding of what engineers and designers do, and how and why they do it. It is only on this basis that the real consequences of the technologies for strategic management can be assessed.

So whilst we concur that codification is important in the development and use of parts of the new electronic toolkit associated with design and engineering routines, this debate ignores the more interesting aspects of innovation associated with the use of the new electronic toolkit; namely, new modes of working, through creative interactions and environments in which designers work together with their computer models, and often in collaboration with users and customers. An underlying assumption in much of the codification argument is that ICTs replace human experience with algorithms embedded in software. But the more interesting develop-ments are occurring in the manipulation of digital symbols and models by highly skilled creative craftspeople, often working in small teams: what McCullogh calls abstracting craft (1996). Rather than replacing trad-itional design skills, the new toolkit complements them in novel and evolving ways.

In this chapter we examine the technologies used in the intensification of innovation with a knowledge management perspective, and consider some of the broader implications of these technologies. The tools have wide implications beyond the parochial concerns of their users or man-agers of innovation. The development and use of these technologies have consequences for our understanding of the management of knowledge and innovation and hence for theories of the firm and of strategic man-agement.[1] Section 2 provides an overview of the role of knowledge in firm behaviour and competitiveness with respect to theories of the firm, evolu-tionary economics and strategic management. Section 3 describes various

ICT-related technologies, and their relation to knowledge management, focusing on technologies used in design. Section 4 assesses the impact and relevance of these new technologies for innovation and Section 5 draws conclusions.

2. KNOWLEDGE, THEORY OF THE FIRM AND STRATEGIC MANAGEMENT

Theoretically and empirically, the creation and use of knowledge takes centre stage in explaining firm behaviour and determining firm competitiveness. This assertion is based on a number of assumptions about firms and their strategic behaviour. We assume that:

1. The way that knowledge is constructed as a resource, and dynamically recreated and used by means of routines and capabilities, is the primary way of maintaining sustainable competitive advantage;
2. Firms are heterogenous in their capabilities and in their capacities to marshal knowledge assets internally, and to aggregate and integrate these assets with external knowledge bases;
3. The specialization of knowledge assets within firms, and their dynamic reconfiguration through internal, market or collaborative means, explains a major element of competitive behaviour.

We draw on a range of theories to support these assumptions. Key amongst these are theories of the firm in the tradition of Edith Penrose, evolutionary economics in the tradition of Nelson and Winter, and strategic management in the tradition of Teece and Pisano. These traditions analyse the firm as bundles of resources, routines and capabilities, and consider their construction, internal configuration and reconstitution as the primary determinant of business competitiveness. These theories emphasize the significance of knowledge, and place the practices that surround the use of knowledge as fundamental elements of firm constructs. For example:

1. Resource-based theories (Barney, 1986; Grant, 1991; Penrose, 1959) consider firms as bundles of assets comprised of both tangible and intangible resources and tacit knowledge. For Penrose, it is the firm's capacity to dynamically adjust its resources that sustains competitiveness;
2. Behavioural theories of the firm (Cyert and March, 1963; March and Simon, 1958) analyse the development of firm-specific routines and the conditions necessary for the production of knowledge;

3. 'Learning' theories (Argyris and Schon, 1978; Brown and Duguid, 2000; Senge, 1993) consider the creation and application of knowledge at various levels, its centrality to organizational performance, its construction at an individual and group level ('communities of practice'), and the ways in which individual learning becomes an organizational property;

4. Evolutionary theory (Nelson and Winter, 1982) identifies the significance of routines as the economic analogues of genes in biology. Routines are the organizational memory for an organization, its repository of knowledge and skills:

5. Dynamic capabilities theory (Teece et al., 1994) encompasses the ability of firms to learn to sense the need to change and then reconfigure internal and external competences to seize opportunities created by rapidly changing environments. In this theory, the essence of the firm is its ability to create, transfer, assemble, integrate and exploit difficult-to-imitate assets, of which knowledge assets are key (Teece, 2002);

6. The concept of absorptive capacities (Cohen and Levinthal, 1990) indicates that a firm's own R&D improves its ability to learn from others. It is essentially a theory of the importance of internal R&D in the integration of external knowledge.

We are not attempting to identify commonalities and synergies between these diverse approaches (although we would enthusiastically encourage any attempts to do so, and particularly welcome efforts to integrate them with aspects of transactions cost economics, (cf. Williamson, 1999), but simply to do three things. First, to draw attention to the ubiquity of notions of knowledge in certain theories of the firm and strategic management. Second, to argue that the construction of resources, routines and capabilities is associated with the creation and use of knowledge. Third, as notions of resources, routines and capabilities are theoretical constructs, difficult to operationalize and examine empirically, we contend that it is valuable to identify the significance of any mechanisms, internal or external to the firm, which can assist the formulation and use of resources, routines and capabilities through the management of knowledge, and can be empirically tested.

Our contention is that the new technologies for innovation provide such a mechanism, but that they do so in highly varied and contingent ways. Although our focus is the introduction and use of specific technologies, we are highly aware that technology and innovation need to be located in particular social and cultural environments that affect their development and use. From the earlier work of David Noble on the development of the automated machine tool (Noble, 1986), to more contemporary analysis of the development of the Linux operating system (Tuomi, 2002), the evidence shows that technology is created and used within the context of strong social, cultural and political influences. Our concern is to identify and reemphasize some of these factors in the introduction of this particular group of new technologies.

3. THE NEW TECHNOLOGIES FOR INNOVATION: THEIR POTENTIAL IMPACT ON KNOWLEDGE MANAGEMENT

Technologies for the new innovation process are emerging from the infrastructure created by what is commonly described under the umbrella of ICTs. At the heart of the new ICTs is increased computational power and faster broadband transmission. The benefits of ICTs are based on large increases in computational speed, processing power and visual display, coupled with cost reductions in equipment.

These technologies are recognized as powerful tools when applied to innovation.[2] The OECD, for example, argues that ICTs are

> . . . a key technology for speeding up the innovation process and reducing cycle times, it has fostered greater networking in the economy, it makes possible faster diffusion of codified knowledge and ideas and it has played an important role in making science more efficient and linking it more closely with business. (OECD, 2000, p. 8)

Thomke (2001) contends that '. . . new technologies such as computer simulation, rapid prototyping, and combinatorial chemistry allow companies to create more learning more rapidly, and that knowledge, in turn can be incorporated in more experiments at less expense' (p. 68).

Table 1.1 lists some of the major technologies used in manufacturing industry, which encompass and are applied to design, production and coordination activities. Of particular recent importance here are the (still emergent) technologies associated with web services. These are based upon shared protocols and standards that allow users to share data and services without requiring translation by human intermediaries (Ismail et al., 2002). The primary focus of this chapter is technologies for design, however the capacity of these technologies to be linked and integrated with production and coordination technologies is critical to their efficacious application.[3]

Research into the diffusion of these technologies shows that they are relatively widely used, across a range of industries and size of firm, with the technologies of production being the most commonly used (Dodgson et al., 2002).

Taking a management of knowledge perspective, the technologies can be assessed according to their capacities to store, search for, connect, represent and create knowledge. The value of these technologies in producing stores or repositories to assist individual and organizational memory is obvious and relatively uncontroversial.[4] They are also useful in searching for, connecting and disseminating sources of knowledge. Internet technologies, including browsers and search engines, help the search process, and hyperlinks assist in the automatic connection of diverse materials. Connections through email and intranets increase the efficiency and speed of sharing documents, diagrams, formulae, symbols and images. Groupware facilitates the communication and sharing of information and knowledge.

The enormous amount of information collected in various databases enables 'data mining' which assists the discovery of commercially valuable knowledge. Data mining is helped by easily used, intuitive navigation tools that can valuably assist in making sense of, and deriving patterns from, the previously unobservable or incomprehensible.[5] The ability to visualize and navigate within datasets, recognizing patterns and finding new relationships has rapidly emerged as a new 'craft' skill (cf. Hutchins, 1996).

As well as web services, the development of e-science platforms is beginning to assist the connectivity of vast, diverse databases, and to enable easy access, data interpretation and knowledge extraction.[6] An example is provided by The Grid in the UK, which is an emergent infrastructure involving universities in every region of the UK, industrial firms, and international collaborators. The UK government is to invest £120 million in The Grid, which it describes as a 'flexible, secure and coordinated resource-sharing infrastructure, based on dynamic collections of computational capacity, data storage facilities, instruments and wide area networks which are as easy to access as electricity in the home' (DTI, 2002). Pilot projects on The Grid include testing, diagnostics, modelling and visualization in aircraft engines, materials design, high performance computing and protein folding.

Technologies such as three and four dimensional CAD (computer aided design) provide a vastly more sophisticated capacity to represent information. They may make complex data, information, perspectives and preferences from diverse groups visible and comprehensible. An example of such use can be provided by the design of a new building. Virtual representation can assist architects in their visualization of the eventual design, and help clarify expectations of the clients (who get a good understanding of what a building will look like), and inform contractors and builders of specifications and requirements (Anumba et al., 2000; Whyte, 2002; Whyte, 2003). Advanced CAD systems – such as originally Dassault's and now IBM's CATIA (Computer Aided Three-dimensional Interactive Application) – is used in the design of products as diverse as helicopters (Sikorsky), automobiles (Rolls-Royce, BMW, Volkswagen), aircraft (Boeing), buildings (Bilbao Guggenheim), shoes (Nike) and home appliances (Braun).

Sophisticated CAD packages, such as CATIA, enable the modelling and engineering of curved surfaces which has been shown to provide additional benefits in bringing more craft back from the designer into the end product. For example, the design of the Bilbao Guggenheim Museum involved more input from Frank Gehry than a traditional process would have been capable of supporting. Gehry is as much sculptor as architect and whilst he hardly cares how a computer works, it was possible to model and engineer his design concepts through many iterations using a toolkit including paper, clay and plasticine models, a scanner adapted from neuroscience, the CATIA package and a rapid prototyping station adapted from the automotive industry. This toolkit enhanced Gehry's role providing more of his 'craft' in the overall solution (Gann, 2000).

Other technologies may be of even greater assistance in the coalescence and creation of knowledge. As Thomke (2001, p. 68) puts it '. . . integrat[ing] new information-based technologies does more than lower costs; it also increases the opportunities for innovation. That is, some technologies can make existing experimental activities more efficient, while others introduce entirely new ways of discovering novel concepts and solutions.'

For example, computer-based simulation has replaced the traditional trial and error approach to many design tasks. Computers can reliably and realistically imitate the behaviour of materials:

> Simulation is the no-risk way of gaining experience before making any investments and the elegant alternative to costly and time-consuming series of tests. The aim is always to analyse, predict, gain experience from the future, and improve. Simulation makes it possible to gain an overview of complex processes . . . It is even possible to conduct experiments by computer which are not feasible in reality, or are too expensive, too dangerous or ethically unacceptable . . . (Miller, 2001, p. 6)

A classic example of the use of simulation techniques is in computer-generated crash tests for automobile designers. This technology, based on CAD programmes, simulates crash behaviour and through repeated redesign stages assists designers in the optimization of their designs. The advantages for the automaker are significant. First, the market demands regular production of new models, and by avoiding the production of prototypes to be tested for safety at early stages in the design process, which can take months, development times are shortened. Second, the expense of crashing physical prototypes is avoided.[7] Third, the quality of data from the simulation is better than that derived from physical prototypes, with resulting improvements for safety.[8]

Other examples of these sorts of technologies include expert systems which may guide good practice and support decisionmaking processes in product development and concurrent engineering. Steinmueller (2000) refers to the way that automated tools are necessary for managing the user's cognitive processes in the use of 'working models', simulations, and large scale computer-supported collaboration. He argues that the extent to which progress is being made in simulation techniques and the construction of virtual models of physical systems is controversial, but nonetheless, there are areas of significant technological importance such as electronics where there exist widely used symbolic systems for representations of designs and specifications. He uses the example of SPICE (Simulation Program Integrated Circuit Emphasis) and other circuit simulation models and symbolic languages for the description of integrated circuit fabrication that allow the reproduction of highly sophisticated working electronic systems and devices (Steinmueller, 2000, p. 372).

Virtual reality technologies, combining supercomputers, high speed Automated Teller Machine (ATM) networks, image generating software and advanced curved screen projection systems allow researchers and designers to visualize complex data sets and enable high performance modelling (Whyte, 2002, 2003). Uses range from pharmaceutical development, where the structure of complicated molecules is visualized to enable more creative and efficient drug design, to the mining industry with the visualization of the properties of ore bodies, enabling economically efficient and environmentally less damaging mining. One of the advantages of these technologies, to be expanded upon below, is the manner in which they enable the virtual proof of concept and enable customer or stakeholder buy-in.

4. ASSESSING THE IMPACT AND RELEVANCE OF THE NEW TECHNOLOGIES FOR INNOVATION

At their core these technologies result from the massive reduction in the cost of storing, transferring and manipulating data and information. The new tools have major implications for the cost and speed of the innovation

Table 1.1   Technologies underpinning the new innovation process[a]

Technologies of design:	Technologies of production:
Computer-aided design	Computer numerical control
Computer-aided engineering	machine tools
Simulation and electronic	Robots
prototyping	Automated transfer systems
Artificial intelligence	Flexible manufacturing cells
Databases and data-mining	Flexible manufacturing systems
Expert systems	Computer aided design/
Electronic exchanges of CAD files	Computer aided manufacture
Optimization tools	(CAD/CAM)
Project extranets	Computer integrated manufacture
Virtual reality	(CIM)
Virtual customer toolkits	Computer integrated production (CIP)
	Lasers for material processing
Technologies of coordination:	High speed machining
Materials requirement	Near net shape technologies
planning (MRP)	Programmable logic controllers
Manufacturing resource	Automated storage/retrieval system
planning (MRPII)	Supervisory control and data
Enterprise resource planning (ERP)	acquisition (SCADA)
Product data management (PDM)	Digital or remote controlled process
Project management systems	plant control
Total quality management (TQM)	Knowledge-based software
Just-in-time delivery systems (JIT)	Rapid protyping
Internet/Intranet/Extranet	
Electronic data interchange (EDI)	
and e-commerce	
Local area networks (LANS)	
Web services	

Notes:
a   This list is based upon current classifications of design, production and coordination technologies. We are working on a new classification which provides a more contemporary taxonomy suitable for analysis of utilization of digital systems and tools for innovation.

process. Where it exists, empirical evidence shows that the technologies can bring substantial cost and speed advantages to their users (Dodgson et al., 2002).

The fact that these technologies have significant consequences for the management of knowledge and for strategic management is beginning to be appreciated within the innovation studies literature (Antonelli and Geuna, 2000; Pavitt, 2002).

However, there is a need for substantially more empirical research, and greater appreciation of the nature of design and prototyping, for their consequences to be properly evaluated.

For example, whether considering the production of knowledge from a theoretical perspective (Gibbons et al., 1994), or the structures and practices that firms and universities use to generate and assimilate knowledge from an empirical perspective (Reger, 1997; Williams, 2002), it is widely accepted that knowledge creation and innovation have become more complex and collaborative. Whilst this is recognized, many current analyses of the impact of the new technologies on innovation within the innovation studies literature fail to address central aspects of these characteristics.

They fail, for example, to place sufficient emphasis on the market-driven factors underlying the complexity and collaborative nature of innovation and knowledge production. The technologies are developed and used primarily in order to deliver better value to customers, through, for example, greater speed and predictability in delivery, reduced costs, improved ability to deal with complexity or customized solutions or the bundling together of products and services. The ability of these technologies to facilitate greater user or consumer input into the innovation process has considerable implications for strategic management.

It has long been recognized that engagement with sophisticated end-users enhances innovative product development (von Hippel, 1988) and can lead to the development of robust design iterations and families of products (Gardiner and Rothwell, 1985). Simulation of the use of products and facilities and working with users to evaluate options as part of the 'design conversation' is becoming a major part of the design and engineering process: from the development of a new washing powder by Proctor & Gamble, through simulation of fire and means of escape on the New York Twin Towers project by Arup, to the design of a new house by Sekisui in Japan. Computer-generated models are becoming more sophisticated, including social and cultural attributes: such as whether and how people leave a building individually or in groups. Some of these technologies are also being developed through competitive collaboration – such as Arup's use and subsequent enhancements of STEPS (Simulation of Transient Evacuation and Pedestrian Movements), and their competitor, Mott MacDonald's software package for fire engineering.

Such levels of involvement and integration of users and consumers through the use of these technologies facilitate industrial specialization and disaggregation. They enable firms to concentrate upon their areas of comparative expertise, safe in the knowledge that the capacity exists to search for and access complementary technologies and services.[9] In this sense, risk is reduced.

There is a need, furthermore, for better appreciation of the role of these tools in facilitating collaboration. Sharing information in a cost effective and prompt manner by means of these technologies is relatively simple: sharing knowledge is considerably more complex. A number of observers of electronically mediated knowledge sharing emphasize the manner in which effective knowledge exchange occurs best amongst groups with some sense of shared identity or trust ('epistemic communities' (Steinmueller, 2000), or 'communities of practice' (Brown and Duguid, 2000), or what Schrage (2000) calls 'shared space'). Salter and Gann (2003) show that although designers are keen users of electronic tools, they rely heavily on close, personal interaction to solve problems, to develop ideas and to assess the quality of their work.

Furthermore, we believe that discussion on the possibilities and limitations of the use of these technologies around the management and economics of knowledge has been somewhat restricted. For example, a key issue in the debate around the creation and use of knowledge is the relationship between its tacit and manifest forms, and much discussion to date on these technologies has focused upon the single and limiting issue of the codification of tacit knowledge.[10]

The value of such technologies in managing existing, codified knowledge is widely acknowledged. As Nonaka et al. (2001) contend, currently ICT is mostly used as a set of tools with which to improve efficiency in combining and disseminating existing information and explicit knowledge. They argue, however, that 'these tools do not offer an integrated and holistic way of dealing with tacit and explicit knowledge in the context of the knowledge economy' (p. 827). In contrast, Antonelli and Geuna (2000) argue that the new technologies enable firms to accumulate tacit knowledge more systematically.

In practice, there are important reflexive and iterative relationships between tacit and explicit knowledge (Nonaka and Takeuchi, 1995). The two forms are not mutually exclusive and neither exists purely by itself. There are additional complications: in our research on mapping and measuring technical excellence in engineering design, with 12 international engineering design firms, we were unable to come to an agreement over the definition of what constituted good design (Gann and Salter, 1999). This was because the types of design activities involved a lot of craft knowledge that was difficult to articulate and codify. However, the group was able to develop a shared understanding of what types of working environments led to better design results.

These issues have significant consequences for our analysis, which argues that although the technologies to which we refer have some potential to codify, represent or make more explicit actions and behaviours that were previously tacit, no matter how automated or codified the technology, there will remain a tacit element to the use of these technologies which may, in the end, provide the defining element of competitive advantage.[11]

We concur with the sentiment of Steinmueller (2000) that individual cognition and social organization are likely to be as significant in the process of knowledge codification as technological issues. However, we believe that it is necessary to move beyond the abstract thesis that the technologies, suitably filtered by social and organizational influences, can lead to codification. Far greater detailed analysis of the actual nature of, and changes occurring in, the design process will produce much richer understanding.[12]

Design needs to be deconstructed so as to ascertain the empirical reality of what engineers and designers actually do, and the changes that are occurring in their roles. Design, for example, is an extraordinarily complex activity, and has been described as a 'social process awash with uncertainty and ambiguity' and as a negotiated trade-off attempting to produce coherence between different participants, with different competencies, skills, responsibilities and interests (Bucciarelli, 2002).

Specifically, if the impact of technologies on knowledge is to be properly understood we believe it is necessary to understand the intricacies and meanings for designers of:

1. Sketching (Bilda and Demirkan, 2003; McGown et al., 1998; Tovey, et al., 2003), and particularly the ways in which sketches are used to communicate, direct and stimulate thought;
2. Visualization (Dahl et al., 2001; Oxman, 2002; Whyte, 2002) is critical during the front end of the design process, in concept design. It can be based on memory (which can be computer-assisted) or imagination (which conceivably cannot be computer-assisted);
3. Language and cultural references (Eckhert and Stacey, 2000), and particularly the ways in which complex concepts are often concisely expressed by references to informal vocabulary or sources of inspiration rather than by explicit language.

It is also necessary to recognize that design is a process with a number of stages, with consequences for the use of the new technologies. To date, most focus has been on the use of the new technologies in the later stages of design, which are information-rich, and intended to embody designs and prepare them for manufacture, rather than the conceptual stages which are typified by vague knowledge and shifting goals (McGown et al., 1998). This distinction is particularly important as it is the early conceptual stages that have the highest impact on quality and costs of future products. It is argued that

70 per cent of product costs are determined within the first fifth of the development process (Romer et al., 2001).

When Arup carried out the engineering design for the Millennium Footbridge across the Thames in London they set out on an adventure to solve a design problem that hitherto had not been addressed – how to produce a very slim structure with the suspension cables running horizontally along the sides of the bridge, rather than vertically above it. The company used some of their most advanced modelling and simulation tools to produce a solution that would not have previously been possible. However, on completion and with a crowd of people on it, the bridge exhibited a form of slow lateral excitation and had to be closed while remedial work was carried out. The problems created by the new possibilities of the design tools had to be solved using a combination of sketching, physical models, remote and face-to-face group interactions and advanced computer modelling. One consequence was that the company developed new mathematical models about the ways in which groups of people walk and the performance of this type of structure. They published these on their website (www.arup.com/millenniumbridge/). The example shows that the use of the new toolkit can lead designers into new terrains and mistakes, or unexpected consequences might ensue. To engineer, after all, is human (Petroski, 1985). But the end result is an innovative structure that enhanced the craft design capabilities of the engineers who built it, whilst also making a wider contribution to knowledge. The development of new knowledge through this design conversation was a much more important outcome of the use of advanced modelling tools than checking code and solving detailed abstract problems.

Furthermore, empirical research to date (Romer et al., 2001; Salter and Gann, 2003) shows that rather than replacing traditional design tools, new electronic media coexist and can complement one another. When it comes, for example, to the use of hand sketches and CAD development in automotive design, what is emerging is a hybrid form of operation (Tovey et al., 2003).

It is also necessary, we believe, in order to understand the significance of these technologies for the management of knowledge and strategic management, to appreciate the central role of prototyping in contemporary business (Schrage, 2000). For Schrage, the new tools are really about creating environments for interaction, a landscape where all the diverse contributors to design meet to help make better choices about uncertain and ambiguous futures. In this sense, good prototypes allow people to experiment with different ideas, stories, models and visions of products. The new tools allow new conversations, the stimulation of new ideas and improvisation around unanticipated ways of creating new value.[13]

5.   CONCLUSION

Most academics studying innovation agree that the influence of the new electronic toolkit will be profound, but that it is still emergent. The contours of use are not well defined. In these circumstances, the real issues are to understand what these tools mean for the organization of design and engineering and the ways in which engineers solve problems: issues such as those raised by Vincenti, including incrementalism, informed guesses, structuring of problems and patterns of interactions of designers as they collaborate face to face. They also, we believe, have consequences for deliberations over the nature of the engineering profession, and its struggle to engage with the nature and integration of design and systems (Williams, 2002). If we are to understand how to manage knowledge in this emerging environment we need to research what this toolkit means for processes and skills, not whether it leads to some abstract substitution and codification of skill.

We do not believe that many of the common assumptions about the use of new electronic tools for innovation hold. For example, we do not think that:

1.   Technology determines work organization, and holds unlimited potential for disembodied automation of the innovation and knowledge management process;
2.   Technology push explains innovations of this nature. We believe that the demand for these technologies derives from market needs to quicken the delivery of new products and services, to integrate the requirements of users into innovation, and to be cost reducing;
3.   New vintages of technology quickly make existing technologies redundant. For example, we do not argue that simulation techniques will replace physical prototyping altogether – the Gehry and Millennium Footbridge examples prove the reverse. They will coexist, and hybrid forms of operating will evolve;
4.   The human element of creativity, problem solving, learning and judgement will be superseded by technology, nor that vast elements of existing tacit knowledge will become easily codified.

We do believe that currently the competitive advantages to be derived by firms from innovation lie in creative leadership in design and development linked with effective integration of other productive functions, the capacity to manage complexity, and in the ability to fully engage the users of innovation in the process of its realization. Our essential argument is that the new technologies for innovation provide new mechanisms for the management of knowledge and innovation and the construction of the resources, routines and capabilities that are the basis of competitiveness. For these technologies to become integrated tools for knowledge management they need to be viewed with a strategic perspective. Research needs to be informed by an understanding of the potential of design and prototyping to be the central element of innovativeness in the sense that has been identified by Schrage (2000). And additionally, they need to inform and facilitate the disaggregation of customer/supplier and networking relationships amongst firms with specialized knowledge assets, and the production of complex products and systems. By understanding the use of the new toolkit as part of the creative motor in the firm, one can then start to see how and where knowledge management might be important.

Whilst there has been some analysis of these issues in the innovation literature, it has to date failed to illuminate the diversity and complexity of empirical practices associated with the use of these technologies. Analysis limited to whether or not the new technologies can codify knowledge ignores the richer questions. Further research is required into the actual technologies that facilitate knowledge storage, search, connectivity, representation and creation; the rich variation in the ways in which designers and innovators use the technologies in, for example, sketching, visualization and language (Whyte, 2002); and the processes by which the technologies are used in various aspects and stages of innovation and design.

It is only on the basis of this research that proper analysis can be undertaken into the ways in which the tools described can assist the construction and application of resources, routines and capabilities. It is by the refinement and empirical testing of categories such as those suggested above that the continuing dialectic between craft and code can be understood and managed.

NOTES

1. If these claims appear outlandish, compare them with Tom Peters' Introduction to Michael Schrage's book on prototyping, where he argues that 'rapid prototyping is the cornerstone, the cultural fountainhead of the innovative enterprise' (Schrage, 2000).
2. We refer here to existing, widely available technology. The capacities of the technologies we discuss will be significantly enhanced with the diffusion of network grids, which create virtual supercomputers of enormous capacity, neural networks, or software designed to mimic human brain activity and learning models, and, eventually, by quantum computing.
3. In discussions of these technologies and their impact upon the management of knowledge, it should be acknowledged that major facilitators of the relationship are technologies that enable *security* in holding and transferring information and knowledge, through, for example, encryption techniques and firewalls (Nonaka and Nishiguchi, 2001), and those that help measure, control and diagnose the results of research.
4. There may, of course, be cognitive problems associated with remembering that information is stored, and technological problems associated with incompatible generations of technological media.

5. For example, DNA micro-array techniques can now test hundreds of thousands of compounds against a target field in a matter of weeks. It previously took months to analyse a few hundred.
6. For example, in the fields of functional genomics and bioinformatics, following the Human Genome Project there are a variety of repositories of knowledge, ranging from basic gene sequences to complex, three dimensional protein structures. Researchers need high level search and knowledge acquisition tools, including computationally intensive simulations, to enable them to study issues like protein folding or receptor docking mechanisms (www.research-councils.ac.uk/escience).
7. Thomke (2001) uses the example of a BMW side-impact crash simulation, where all 91 virtual crashes cost less than the $300 000 cost of a physical test.
8. Engineers at BMW dismantled a car into 60 000 pieces by means of computation. Using precisely defined accident data, the software calculates the mechanical forces which adjacent elements exert on each one of these finite elements. With knowledge of the material properties of the element, engineers can visualize the progress of the deformation process at the intervals of one millionth of a second (Fraunhofer Magazine, 2002, p. 23).
9. The recent McKinsey analysis of web services argues that their likely result is: 'increased fragmentation of value chains and industries as well as more narrowly focused companies' (Ismail et al., 2002).
10. See the special edition of *Research Policy* 'Codification of Knowledge: Empirical Evidence and Policy Recommendations', 30 (9), 2001, and *Industrial and Corporate Change*, 9 (2), 2000.
11. We are currently exploring the role these technologies play in relation to tacit and explicit knowledge. We agree with Cook and Brown (1999) that these are distinct concepts and that it is not necessarily possible to use one form as an aid to acquiring the other.
12. It is notable that in neither of the journal special editions on the codification of knowledge cited in note 10 was there any reference to the journal *Design Studies*, which is a major venue for discussing the nature of design and the work of designers and the ways in which they are changing as a result of the kinds of technologies discussed in those special editions.
13. The range of insights from this literature will be merged with those from the innovation and knowledge management perspective in a forthcoming book by the present authors, scheduled to be published by Oxford University Press in 2005.

REFERENCES

Anumba, C. J., N. M. Bouchlaghem, J. K. Whyte and A. Duke (2000), 'Perspectives on a Shared Construction Project Model', *International Journal of Cooperative Information Systems*, 9 (3), 283–313.
Antonelli, C. and A. Geuna (2000), 'Information and Communication Technologies and the Production, Distribution and use of Knowledge', *International Journal of Technology Management*, 20 (1/2), 72–105.
Argyris, C. and D. Schon (1978), *Organizational Learning: Theory, Method and Practice*, London: Addison Wesley.
Barney, J. (1986), 'Strategic Factor Markets: Expectations, Luck, and Business Strategy', *Management Science*, 32 (10), 1231–41.
Bilda, Z. and H. Demirkan (2003), 'An Insight on Designers' Sketching Activities in Traditional versus Digital Media', *Design Studies*, 24 (1), 27–50.
Brown, J. S. and P. Duguid (2000), *The Social Life of Information*, Boston, MA: Harvard Business School Press.

Bucciarelli, L. (2002), 'Between Thought and Object in Engineering Design', *Design Studies*, **23** (3), 219–31.

Cohen, W. and D. Levinthal (1990), 'Absorptive Capacity: A New Perspective on Learning and Innovation', *Administrative Science Quarterly*, **35** (1), 128–52.

Cook, S. D. N. and Brown, J. S. (1999), 'Bridging Epistemologies: The Generative Dance between Organizational Knowledge and Organizational Knowing', *Organization Science*, **10** (4), 381–400.

Cyert, R. and J. March (1963), *The Behavioural Theory of the Firm*, Englewood Cliffs, NJ: Prentice-Hall.

Dahl, D., A. Chattopadhyay and G. Gorn (2001), 'The Importance of Visualization in Concept Design', *Design Studies*, **22** (1), 5–26.

Dodgson, M., D. M. Gann and A. Salter (2002), 'The Intensification of Innovation', *International Journal of Innovation Management*, **6** (1), 53–84.

Dodgson, M., D. M. Gann and A. Salter (forthcoming), *Think, Play, Do: Technology and Innovation in the Emerging Innovation Process*, Oxford: Oxford University Press.

Department of Trade and Industry (DTI) (2002), 'E-Science: Building a Global Grid', report, London: DTI.

Eckert, C. and M. Stacey (2000), 'Sources of Inspiration: A Language of Design', *Design Studies*, **21** (5), 523–38.

Fraunhofer Magazine (2002) 'Simulation at the Touch of a Button', *Fraunhofer Magazine*, **1**, 22–3.

Gann, D. M. (2000), *Building Innovation: Complex Constructs in a Changing World*, London: Thomas Telford Press.

Gann, D. and A. Salter (1999), 'Discussion paper on CIRIA Technical Excellence Benchmarking Group', unpublished manuscript, Science and Technology Policy Research (SPRU), University of Sussex, Falmer.

Gardiner, P. and R. Rothwell (1985), 'Tough Customers: Good Designs', *Design Studies*, **6** (1), 7–17.

Gibbons, M., C. Limoges, H. Nowotny, S. Schwartzmann, P. Scott and M. Trow (1994), *The New Production of Knowledge: The Dynamics of Science and Research in Contemporary Societies*, London: Sage.

Grant, R. M. (1991), *Contemporary Strategy Analysis*, Cambridge, MA: Blackwell Business.

Hutchins, E. (1996), *Cognition in the Wild*, Cambridge, MA: MIT Press.

Ismail, A., S. Patil and S. Saigal (2002), 'When Computers Learn to Talk: a Web Service Primer', *McKinsey Quarterly*, special edition (2), 70–7.

March, J. and H. Simon (1958), *Organizations*, New York: Wiley.

McCullogh, M. (1996), *Abstracting Craft: The Practiced Digital Hand*, Cambridge, MA: MIT Press.

McGown, A., G. Green and P. Rodgers (1998), 'Visible Ideas: Information Patterns of Conceptual Sketch Activity', *Design Studies*, **19** (4), 431–53.

Miller, F. (2001), 'Simulation – Can we Compute the Future?', *Fraunhofer Magazine*, **1**, 6–11.

Nelson, R. and S. Winter (1982), *An Evolutionary Theory of Economic Change*, Cambridge, MA: Belknap Press.

Noble, D. F. (1986), *Forces of Production: A Social History of Industrial Automation*, New York: Oxford University Press.

Nonaka, I. and T. Nishiguchi (2001), *Knowledge Emergence: Social, Technical and Evolutionary Dimensions of Knowledge Creation*, Oxford: Oxford University Press.

Nonaka, I. and H. Takeuchi (1995), *The Knowledge-Creating Company: How Japanese Companies Create the Dynamics of Innovation*, New York: Oxford University Press.

Organisation for Economic Co-operation and Development (OECD) (2000), *A New Economy? The Changing Role of Innovation and Information Technology in Growth*, Paris: OECD.

Oxman, R. (2002), 'The Thinking Eye: Visual Re-cognition in Design Emergence', *Design Studies*, **23** (2), 135–64.

Pavitt, K. (2002), Systems Integrators as "Post-Industrial" Firms, Brighton: SPRU.

Penrose, E. (1959), *The Theory of the Growth of the Firm*, Oxford: Oxford University Press.

Petroski, H. (1985), *To Engineer Is Human – The Role of Failure in Successful Design*, New York: St Martin's Press.

Reger, G. (1997), 'Benchmarking the Internationalization and Co-ordination of R&D of Western European and Japanese Multi-national Corporations', *International Journal of Innovation Management*, **1** (3), 299–331.

Romer, A., G. Weisshahn, U. Lindemann and W. Hacker (2001), 'Effort-saving Product Representations in Design – Results of a Questionnaire Survey', *Design Studies*, **22** (6), 473–91.

Salter, A. and D. M. Gann (2003), 'Sources of Ideas for Innovation in Engineering Design', *Research Policy*, **32** (8), 1309–24.

Schrage, M. (2000), *Serious Play: How the World's Best Companies Simulate to Innovate*, Boston, MA: Harvard Business School Press.

Senge, P. M. (1993), *The Fifth Discipline: The Art and Practice of the Learning Organization*, London: Century Business.

Steinmueller, E. (2000), 'Will New Information and Communication Technologies Improve the 'Codification' of Knowledge', *Industrial and Corporate Change*, **9** (2), 361–76.

Teece, D. (2002), 'Dynamic Capabilities', in W. Lazonick (ed.), *The IEBM Handbook of Economics*, London: Thomson.

Teece, D., G. Pisano and A. Shuen (1994), 'Dynamic Capabilities and Strategic Management', *Strategic Management Journal*, **18** (7), 509–33.

Thomke, S. (2001), 'Enlightened Experimentation: The New Imperative for Innovation', *Harvard Business Review*, **79** (2), 67–75.

Tovey, M., S. Porter and R. Newman (2003), 'Sketching, Concept Development and Automotive Design', *Design Studies*, **24** (2), 135–53.

Tuomi, I. (2002), *Networks of Innovation: Change and Meaning in the Age of the Internet*, New York: Oxford University Press.

Vincenti, W. G. (1990), *What Engineers Know and How They Know It*, Baltimore, MD and London: Johns Hopkins University Press.

von Hippel, E. (1988), *The Sources of Innovation*, Oxford: Oxford University Press.

Whyte, J. K. (2002), *Virtual Reality in the Built Environment*, London: Architectural Press.

Whyte, J. K. (2003), 'Industrial Applications of Virtual Reality in Architecture and Construction', *Electronic Journal of Information Technology in Construction*, special issue on virtual reality technology in architecture and construction, **8**, 43–50.

Williams, R. (2002), *Retooling: a Historian Confronts Technological Change*, Cambridge, MA: MIT Press.

Williamson, O. E. (1999), 'Strategy Research: Governance and Competence Perspectives', *Strategic Management Journal*, **20** (12), 1087–108.

2. The economics of governance: the role of localized knowledge in the interdependence among transaction, coordination and production

Cristiano Antonelli

1. INTRODUCTION

Transaction costs economics has made possible significant progress in the economic analysis of the firm. The continual process of implementation and redefinition of the original framework put forward by Ronald Coase and Oliver Williamson and the contributions of the resource-based theory of the firm have paved the way to a broader approach: the economics of governance.

In transaction costs economics the firm is viewed as a bundle of activities selected according to the relative costs of transaction and coordination. Internalization is decided when the costs of using the markets are higher than the costs of coordinating production internally. The basic choice is whether to buy a given component or other intermediary inputs or to make them. The decision is taken in a static context where coordination and transaction costs are given and depend upon exogenous factors. The role of competence and knowledge is not considered.

An alternative view of the firm has been elaborated by the resource-based theory of the firm. The resource-based theory of the firm has emerged as a consistent body of literature centred upon the key role of the firm in the accumulation and generation of technological knowledge and competence and its transformation into technological and organizational innovations (Foss, 1997; Penrose, 1959).

In the resource-based theory of the firm little attention is paid to understanding the role of coordination costs in limiting the size of the firm and to the constraints and opportunities of the marketplace as an alternative mechanism of governance.

The analysis of coordination and transaction-specific activities cannot be conducted in isolation with respect to the choices and the characteristics of the production process and the markets for products and intermediary inputs. The decisions of inclusion and exclusion of each specific segment of the production process can be assessed only when coordination and transaction are viewed as the result of well specified forms of economic activity characterized by their own specific form of competence and organizational knowledge.

This makes it possible to move from transaction costs economics towards a broader economics of governance approach. The object of analysis in the economics of governance approach is the organization of the firm with a special emphasis upon the localized process of accumulation of technological and organizational knowledge and the introduction of both technological and organizational innovations.

The rest of the chapter is structured as follows. In Section 2 the comparative assessment of the elements of strength and weakness of transaction cost theory and of the resource-based theory of the firm is elaborated as a step towards an integrated economics of governance. Section 3 discusses the interdependence between production, transaction and coordination and provides an analytical model, which is subsequently applied, in Section 4, to grasp the complexities of interdependence in a dynamic context. The conclusions summarize the argument and put it in perspective.

2.   TOWARDS AN ECONOMICS OF GOVERNANCE

Two different approaches confront each other in the theory of the firm: transaction cost economics and the resource-based theory.[1] A comparative analysis makes it possible to stress their relative advantages as well as their weaknesses. In so doing it provides the elements to elaborate an integrated approach.

2.1   From Transaction Costs Economics to the Economics of Governance

Transaction cost economics is the result of an incremental process of extension and implementation of the framework first elaborated by Ronald Coase. Oliver Williamson provided an operational context which proved to be extremely fertile.

The unit of analysis here is the transaction. The firm is viewed as a nexus of contracts and a portfolio of given production functions which coexist within the same organization according to the trade-off between coordination and transaction costs. The choice whether to include or exclude a given production process within the borders of the firm depends upon the levels of coordination and transaction costs respectively. When the costs of internal coordination are higher than the costs of using the market, a transaction takes place and that production function remains outside the borders of the firm. Inclusion takes place when the costs of internal coordination are lower than the costs of using the market (Williamson, 1975, 1985, 1990, 1996).

The coordination of diverse activities entails specific costs associated with the need to control the actual performance of the tasks assigned to the agents and to monitor their efficiency. Coordination costs are specific information costs stemming from the bounded rationality and limited knowledge of managers (Alchian and Demsetz, 1972; Simon, 1947, 1982).

Transaction costs depend upon given technological features such as the asset specificity and the frequency of exchanges, the characteristics of the marketplace in terms of transparency, common trust and actual enforcement conditions of obligations in contracts; hence institutional reliability. The levels of transaction costs mainly consist in the costs of the resources that are necessary to search for possible suppliers of specific components and activities, the assessment of their quality, price and delivery conditions, the costs of designing effective contracts with the prospective suppliers and enforcing them. Transaction costs as well are expressions of bounded rationality and limited knowledge, but they concern the perspective external suppliers, rather than internal agents.[2]

In transaction costs economics neither transaction nor coordination are viewed as activities, but solely as costs: there is no analysis of the efficiency of the activities which are put in place in order to perform the required coordination and transaction. There is no analysis of the knowledge and the competence necessary to coordinate and use the markets respectively and hence little room is left for understanding the process of accumulation of new organizational knowledge and the introduction of organizational innovations. By the same token, the technology of the production process is considered as given and exogenous. In transaction costs economics the firm does not consider the issues of the choice among technologies and even less attention is paid to the governance of the accumulation of new knowledge and the introduction of new technologies. The interdependence between technological choices and organizational ones is not considered.

The poor attention paid by transaction costs economics to the conditions and the dynamics of the accumulation and generation of new knowledge and competence is a major weakness. Knowledge and competence applied to the manufacturing processes as well as to the management of the internal coordination and to the procedures and the skills that are necessary to use the markets, are key to understanding the firm. A clear understanding of the role of technological and organizational knowledge in the theory of the firm is provided by the resource-based approach.

2.2   From the Resource-based Theory of the Firm to the Economics of Governance

The resource-based theory provides a distinctive and yet complementary approach to analysing the firm. The emphasis here is put on the process by means of which the firm is able to introduce technological and organizational innovations (Penrose, 1959). The firm is viewed as the locus where technological and organizational knowledge is generated by means of the integration of learning processes and formal research and development activities. The firm is considered in this approach primarily as a depository and a generator of competence (Foss, 1997, 1998; Foss and Mahnke, 2000).

The resource-based theory of the firm has grown as a development and an application of the economics of learning. The enquiry into the dynamics and the characteristics of learning processes, such as learning by doing and learning by using, and their relevance in explaining technological change, has led to the identification of the firm as the primary locus of the generation and valorization of knowledge immediately relevant for economic action, at least in market economies (Arrow, 1962a; Lamberton, 1971; Loasby, 1999).

In the resource-based theory of the firm, the generation of technological knowledge is regarded as the distinctive feature of the firm. The firm does not coincide with the production function and cannot be reduced to a production function because its essential role is the accumulation of competence, technological and organizational knowledge and the eventual introduction of technological and organizational innovations. From this viewpoint the firm precedes the production function: the technology is in fact the result of the accumulation of knowledge and its application to a specific economic activity. Technological knowledge can be considered the primary output of the firm and in turn an intermediary input. The choice whether to sell it or to use and make with it is especially relevant.

In the resource-based theory, the firm cannot be viewed only as a nexus of contracts: the specificity of the production process and the characteristics of the products are a consequence of the process of generation of technological and organizational knowledge. Hence the firm, in the resource-based theory, is much more than a nexus of contracts: it is primarily a mechanism for the production of knowledge.

The resource-based theory of the firm however, has paid little attention to understanding the role of organizational factors in shaping the accumulation and generation of new knowledge. Specifically, the resource-based theory of the firm has not elaborated a full understanding of the constraints, in terms of both rate and direction, on the dynamics of learning, that arise from the costs of using the hierarchies and the markets respectively. Organizational factors shape the valorization of the knowledge accumulated by means of the learning processes and constrain the direction as well as the rate of learning.

The blending between the resource-based theory of the firm and transaction cost economics into a fully articulated economics of governance seems a necessary step to appreciate the key role of localized technological and organizational knowledge in shaping the growth of the firm.

3. THE GOVERNANCE SYSTEM: A MODEL OF INTERDEPENDENCE BETWEEN PRODUCTION, TRANSACTION, COORDINATION AND KNOWLEDGE GENERATION

3.1 General Considerations

The integration between transaction cost economics and the resource-based theory of the firm provides major opportunities for implementing a broader economics of governance. Important complementarities are found when an effort is made to understand the role of competence and knowledge in the definition of the borders of the firm, under the constraint of the resources that are necessary to coordinate the diverse activities retained within its borders. The generation of knowledge is the primary role of the firm but under the constraint of governance costs.

The integration of transaction costs economics and the resource-based theory is possible when attention is focused upon the interdependence between the decisionmaking in the manufacturing activities and in the coordination and transaction ones. In such an approach competence is the basic factor in performing the full range of activities that are necessary to understand the firm. The understanding of the factors affecting the choice between inclusion and exclusion, including the costs of using respectively the markets and the internal hierarchies, is a basic ingredient in a theory of the firm which no longer coincides with the textbook production function.

In the economics of governance the definition of the borders of the firm and the choice between exclusion and inclusion are the result of a broad range of dynamic factors. The assessment of the inclusion/exclusion choice includes the efficiency of the internal manufacturing of the components with respect to their market prices, as well as the competence of the firm in performing transaction and coordination activities respectively. The characteristics of the process of accumulation of technological and organizational knowledge and of the endogenous introduction of new technologies and innovations in the governance activities that are necessary to perform transaction and coordination influence the inclusion/exclusion decision-making, as well as all innovations in the production process.

The economics of governance benefits from the resource-based theory of the firm in expanding the scope of transaction costs economics so as to include the analysis of: (1) the accumulation of competence and knowledge; (2) the introduction and selection of technological and organizational innovations; and (3) their effects on the design of the portfolio of activities which are sorted to be respectively included within the firm and assigned to transactions in the marketplace (Chandler et al., 1998; Penrose, 1959).

The understanding of the overlapping between production theory, economics of innovation and economics of knowledge makes it possible to provide an integrated analytical framework which is able to study the broad range of factors that affect the governance of the firm viewed not just as a nexus of contracts, but rather as a selective and selected combination of complementary activities based upon the capability to accumulate competence and knowledge.

In the economics of governance, the firm is a bundle of activities selected under the constraint of technological, organizational and market factors. No factor can be isolated: the actual size of the firm and its structure can be understood only when the three classes of factors are analysed in close conjunction and an effort is made to appreciate their interdependence.

Specifically within the borders of the firm we can identify production activities, a coordination activity and a transaction activity. The implementation of all activities implies appropriate levels of knowledge and competence and hence of efficiency. The introduction of organizational innovations in coordination and transaction activities and of technological innovations in production, in turn leads to increased efficiency.

The coordination activity provides the management, monitoring and assessment of the relations between the indivisible modules that are retained within the borders of the corporation. The transaction activity consists in the use of the markets for the provision of intermediary inputs.

The borders of the firm are assessed according to the costs of intermediary products internally manufactured relative to the costs of external inputs. The choice between the exclusion and the inclusion of each input is influenced by an array of factors that are strongly interdependent in assessing the size of the portfolio of activities performed within the borders of each firm. The understanding of such interdependence makes possible important progress in the theory of the firm.

Firms select the mix of internal and external products and services according to the combined costs of production and coordination on the one hand and the combined costs of purchasing and using the markets on the other. Coordination activities cannot be separated from the firm's own internal manufacturing of the products and services. By the same token transaction activities cannot be separated from the actual use of the market as an alternative means of procuring or selling some products.

In so doing some substitution takes place. Neither coordination nor transaction activities however can be cancelled. A notion of partial substitutability between coordination and transaction activities emerges. The choice between coordination and transaction, and hence between inclusion and exclusion, can take place, but only up to a point. The traditional analysis of complementary substitutability between production factors, familiar to the theory of production, applies also to the analysis of the governance of firms. This notion of partial substitutability between coordination and transaction activities makes it possible to explore a wide range of mixed governance structures where varying mixes of transaction, production and coordination activities are at work. In so doing the key role of localized technological and organizational knowledge can be fully appreciated.

The cost of internal inputs depends upon the sheer cost of the production process of each activity and the costs of their coordination. The cost of external inputs depends upon their market price and the costs of their procurement in using the markets.[3]

These decisions cannot however be taken without a clear assessment of the costs associated with inclusion and exclusion respectively.

Both coordination and transaction activities are resource consuming. Dedicated inputs are necessary to perform the coordination and transaction activities. The usual relationship between inputs and outputs applies. The efficiency of the coordination and transaction activities is determined by the competence accumulated and the organizational knowledge available to each firm. Higher levels of organizational competence may eventually lead to the introduction of organizational innovations which in turn make it possible to improve the efficiency of both the coordination and the transaction activities (Argyres, 1995).

Here, the interdependence between the factors becomes evident. At each point in time, for given levels of competence in transacting, the adoption of a technology may be influenced by the levels of the transactions costs that are associated with the asset specificity and the frequency of the transactions that characterize it. With different levels of competence however the firm may select other rival technologies. In this approach the technology of each production process is the result of the innovative choice of the firm itself: the characteristics of each technology are not given and exogenous, but are the result of the innovation and the related accumulation of knowledge and competence within the firm itself. Here it seems clear that the conditions of the coordination and transaction activities directly affect the process of generation and use of technological knowledge and eventually the design and the specific introduction of the new technologies (Loasby, 1999; Nooteboom, 2000; Teece, 2000).

The blending of transaction costs economics and the resource-based theory makes it possible to understand the constraints and the limitations that the costs of using the hierarchies and the markets respectively exert upon the accumulation and generation of new knowledge. The firm itself can be regarded as an island of coordination procedures that facilitate the accumulation of knowledge. The Coase-Williamson argument, much applied to the choice between coordination and transaction in the organization of the economic activity, can now be stretched and elaborated so as to understand the characteristics and the effects of the firm as the fabric of localized technological knowledge (Antonelli, 1999a, 1999b, 2001).

In this approach technological knowledge is made possible by the continual efforts of accumulation of competence and technological knowledge based upon localized learning processes and the eventual introduction of innovations by agents rooted in a well defined set of scientific, technical, geographic, economic and commercial circumstances, constrained by substantial irreversibility of both tangible and intangible production factors including reputation and communication channels with customers and suppliers, and yet able to react with creativity to the emerging mismatch between expected and actual market conditions.[4]

3.2 The Model

In standard microeconomics the firm coincides with the production function. In transaction costs economics, coordination and transaction costs define the borders of the firm, but no analysis is provided on the activities that are necessary to perform these functions, the role of competence and knowledge, both in the organization and in the production and their interdependence. On the contrary, in the resource-based theory of the firm, learning generates knowledge and knowledge makes growth possible, but little attention is paid to the constraints and limitations of organizational factors.

In the economics of governance, the output of each firm is the result of the combination between internal and external inputs, respectively manufactured, managed, selected, monitored and purchased by means of dedicated activities. Activities in turn are shaped and characterized in terms of competence and dynamic efficiency.

The firm is viewed as a microsystem where many interdependent learning activities are at work and influence each other. The governance choice is made according to the costs of external inputs and internal ones. These however are determined by the efficiency of the activities that are necessary to produce them.

The governance of the firm can be viewed as the selection of the combination between bundles of production and organizational activities, rather than goods: the selective procurement of external inputs and the production and coordination of internal ones.[5]

A simple governance system of five equations accommodates the analysis elaborated so far. The working of the firm can be grasped by means of a corporate function and a production function where standard substitution takes place and a transaction and coordination activity characterized by fixed coefficients. Each is qualified by the key role of knowledge and competence modelled as a shift parameter. A standard cost function completes the set of constraints that make it possible to analyse the behaviour of the firm. Formally we see the following:

(1) $$Y = A_1(t) ((TRA)^\alpha, (CO)^\beta)$$
(2) $$TRA = (A_2)(t) (EXTERNAL, R)$$
(3) $$CO = (A_3)(t) (INTERNAL, R)$$
(4) $$INTERNAL = (A_4)(t) (K^a, L^b)$$
(5) $$C = p\, EXTERNAL + u\, R + r\, K + w\, L$$

where Y denotes the output levels that are obtained by means of a corporate governance function characterized by a general level of competence $A_1(t)$ that can increase in time and provide the combination of inputs that are either (EXTERNAL) purchased in the marketplace by means of transaction activities (TRA), or internal, i.e. manufactured internally – by means of a standard production function based upon capital (K) and labour (L) inputs as well as specific technological knowledge (A_4) which increases over time because of learning processes and dedicated research activities – and managed by means of coordination activities (CO).

Coordination activities are the product of the organizational resources (R) that perform the specific task of coordinating the inputs produced internally by means of the production function. Coordination activities moreover are characterized by some dedicated level of competence and organizational knowledge (A_3) that is allowed to change over time because of learning and dedicated research activities.[6] Transaction activities are also the output of organizational resources (R) that perform all the clerical tasks that are necessary to purchase the external inputs in the marketplace. Transaction activities in turn are characterized by a specific and dedicated level of competence and organizational knowledge (A_2) that changes over time because of learning processes and dedicated research activities.

For both activities, a fixed coefficient between the amount of internal and external inputs respectively and the organizational resources (R) that are necessary to perform the coordination and transaction activities is given. It may change over time according to the value of the specific shift parameter that measures the rates of accumulation of dedicated knowledge in each activity and to the informational conditions of hierarchies and markets.[7]

The governance function can be characterized by returns to scale that can be increasing or decreasing according to the parameter α and β. The production function in turn can exhibit increasing or decreasing returns to scale according to the value of the parameters a and b.

Next to the governance function there is a general cost function where the costs of the external inputs that enter the transaction activities (p) and the unit costs (u) of the organizational resources (R) that enter both the transaction and coordination activities respectively are considered together with the unit cost of capital (r) and labor (w).

The working of the governance system is quite simple. For given market prices of the output, the firm will select not only the levels of output but also the portfolio of activities according to: (1) the efficiency of the production process; (2) the effects of increasing returns in production; (3) the efficiency of the corporate function; (4) the effects of increasing returns in the corporate function; (5) the efficiency in the transaction activities; and (6) the competence and hence efficiency in coordination activities.

The firm will rely more on external rather than internal inputs when the production function is characterized by a relative inefficiency with respect to other suppliers or when decreasing returns affect its average manufacturing costs, when coordination activities are less effective than transaction activities and hence coordination costs are larger than transaction costs.

The details of the production process, such as the efficiency of the internal production process and the extent to which increasing and decreasing returns are at work, can be assessed with respect to the prices of the products in the markets. The levels of transaction costs, as determined by the dedicated competence of each firm in using the markets, interact both with the comparative costs of the products manufactured internally with respect to their market prices, and the levels of efficiency of the coordination function.[8]

The governance choices are made under the influence and the effects of all the factors that have been considered so far. The quality of the markets, both from an informational and a competitive viewpoint, the characteristics of the products and especially their novelty, the features of the production process both with respect to the levels of asset specificity and to the costs of production, the levels of technological advances in manufacturing, with respect to competitors, and the levels of competence in performing coordination and transaction activities respectively are interdependent factors which influence each other and which cannot be separated and isolated in assessing the governance choice of the firms.

All changes in the levels of competence and in the knowledge base of the firm are likely to affect not only its conduct, but also its structure. The increase in the general knowledge base at the corporate level (A_1) as well as the generation of new production knowledge (A_4) above the average of competitors will favour the expansion of the borders of the firm with processes of vertical integration, diversification and multinational growth. This is also the case when the firm is able to increase its coordination knowledge base (A_3). In contrast, the increase of transaction knowledge is likely to push towards the selection of activities retained within its borders. The dynamics of the knowledge base becomes the central issue in assessing the evolution of the corporation (Loasby, 2002).

4. THE ROLE OF LOCALIZED KNOWLEDGE: DYNAMIC IMPLICATIONS

The model elaborated so far to handle the analysis of the interdependence between production, transaction and coordination activities, is a first result of the attempt to merge transaction costs economics with the resource-based theory of the firm, still in a static context, yet it has many important dynamic implications.

The focus upon transaction and coordination viewed primarily as activities, which entail specific competencies and dedicated levels of organizational knowledge, rather than sheer costs, has in fact direct and relevant consequences in dynamic terms. Here the variety of firms and their localized endowment of competencies and experience, built by means of learning processes, matter.

The firm is no longer viewed as a representative agent. The specific characteristics of the firm need to be investigated and assessed both with respect to the organizational processes and with respect to the production processes. The analysis of production and organization cannot be separated.

The corporation is a resource pool designed and managed so as to implement the opportunities for the accumulation of both new technological and organizational knowledge. The rates of technological and organizational learning influence each other in shaping the dynamics of the firm, the evolving composition of the collection of activities that are retained within its borders and ultimately its growth (Chandler et al., 1998; Teece, 2000).

The notions of localized technological knowledge and localized technological change stress the relevance of the learning processes circumscribed in the specific and idiosyncratic locations within technical, organizational, product and geographical spaces, of each firm at each point in time. The learning processes in such locations are the basic conditions for the accumulation of experience and the eventual generation of both competence and tacit knowledge. On these bases in turn the firm is able to acquire other forms of knowledge, respectively external codified and tacit knowledge and to implement the internal tacit knowledge with research and development activities. In this approach, the firm is primarily defined as a bundle of activities that are complementary with respect to the generation of knowledge and competence (Antonelli, 1999a, 2001).

The characteristics of the process of accumulation of competence, of the generation of technological knowledge and of the introduction of technological and organizational innovations, are key factors in understanding the firm. Parallel to knowledge, competence is a central ingredient. Competence is defined in terms of problem solving capabilities and makes it possible for the firm not only to know how, but also to know where, to know when, and to know what to produce, to sell, to buy. Competence and knowledge apply to the full set of activities: production activities, transaction activities and coordination activities (Nooteboom, 2000).

The dynamics of the firm is shaped by the dynamic interdependence among the accumulation of localized knowledge and competence respectively in coordination, transaction and production (Chandler, 1962, 1977, 1990).

The accumulation of experience and competence in the production process, out of learning processes, leads to more efficient production processes. The costs of internal production are lower than the market prices for the same goods even in competitive markets. The firm internalizes that production even if transaction costs are low and coordination costs are high: production costs matter and interact with the localized organizational decisionmaking.

By the same token all learning in coordination is likely to increase the stock of dedicated organizational knowledge and hence to increase the efficiency of the firm in performing coordination activities. The larger the competence in coordination, the larger the portfolio of activities which can be retained within the borders of the firm. Firms grow into large, diversified, integrated and possibly multinational corporations when coordination competencies are large.

The introduction of an array of innovations in coordination activities, such as the multidivisional form, the matrix structure and in-house outsourcing have made it possible to reduce coordination costs (Bonazzi and Antonelli, 2003; Chandler, 1962, 1977, 1990).

The introduction of major technological innovations, such as new information and communication technologies, has important implications in terms of organizational innovations. Information and communication technologies have made it possible to reduce information asymmetries and hence coordination costs. Similar effects however have been observed in transaction costs: e-commerce and especially e-markets seem to make possible relevant reductions in the costs of transactions (Antonelli, 1988).

Learning in transaction increases the competence of the firm in using the markets and hence reduces the levels of transaction costs with the ultimate effect, ceteris paribus, of pushing the firm to reduce the number of activities retained within its borders. Firms able to elaborate a distinctive competence in dealing with market transactions shrink the size of their portfolios of activities conducted internally but, can extend the scope of their operation as intermediary (Spulber, 1999).

Decreasing returns in the corporate function can become a major obstacle for the firm to benefit from the accumulation of technological knowledge and prevent the successful introduction of technological innovations. Organization costs limit the growth of the firm, when it is based only upon the generation of technological knowledge – or increasing returns in manufacturing – that is not paralleled by the accumulation of organizational knowledge (Arrow, 1974).

Organization costs matter when there is a case of diffusion of new rival technologies and of technological variety at large. Here the selective adoption of one technology instead of another may be influenced by the levels of transaction costs in the marketplace. Transaction costs, in other words, influence the technology rather than being determined by the technology. By the same token coordination costs can affect technological choices.

In the governance economics context of analysis a new area of analysis emerges, one where the governance choice concerns also the markets for outputs, rather than solely the markets for inputs. The firm in fact considers not only the possibility of making or buying a specific component or stage of the production process, but also whether to sell its products in the intermediary markets or to the final ones. Needless to say the stages of the intermediary markets in which to sell are also a matter of choice and assessment. The firm can decide whether to integrate and diversify downward, as well as upward. In this context the firm can also make the choice to sell and eventually to buy again at a later stage of the production process. Here the firm selects the stages of complex and interdependent production processes, which can be internalized, and the stages to externalize, but retains the control of the overall production process articulated in sequential steps. The market and the organization become interdependent. The firm can be at the same time the vendor of a product and the buyer at a later stage. The firm can buy back the full amount of the goods produced with its own original inputs or only a part. The borders between the firm and the markets become more and more flexible and subject to continual redefinition.

The analysis developed so far has important applications to understanding the conduct of the innovative firm when the stock of technological knowledge accumulated within each firm and the competence built by means of learning processes and formal research and development activities is considered an output per se, rather than an input for the subsequent production of goods and services in the markets for technological knowledge. Now the choice whether to make or to buy is integrated by the choice whether to sell or to make. Specifically firms assess both whether to produce internally all the knowledge that is necessary for the introduction of new technology or purchase it in the markets for external knowledge, and whether to sell the knowledge in the markets for knowledge or to use it to make other products.

The use of the marketplace to exchange technological knowledge is more and more common. Technological knowledge can be fully generated internally or partly purchased in the markets for knowledge: external knowledge can be an intermediary input for the production of other knowledge.[9]

Technological knowledge can be sold with varying levels of embodiment into other goods and services. Technological knowledge can be sold as an intangible good, more or less associated with other services such as the assistance of the vendors to the customers. Technological knowledge can be sold embodied at an early stage of a broader production process, or embodied in products that are manufactured at other stages further down in the general production process leading to the products actually purchased by the final consumer: the household (Arora, et al., 2001; Guilhon, 2001).

Knowledge transaction costs, i.e. the costs of using the markets for knowledge, play a key role in this context. In turn, knowledge transaction costs are affected by the characteristics of knowledge, such as appropriability, cumulability, complementarity, fungibility and stickiness. Knowledge transaction costs play a major role in understanding the architectural design of the firm and the combination of activities retained within its borders. Let us analyse them in turn (Arrow, 1962b, 1969).

With low levels of knowledge appropriability and hence high risks of opportunism and dissipation of the rents associated with knowledge, knowledge transaction costs are very high and firms cannot rely on the marketplace to valorize their intangible outputs. The embodiment of technological knowledge into new products and their eventual sale in the marketplace becomes necessary (Antonelli, 2001; Teece, 1986, 2000).

The quasi-private good nature of technological knowledge as a matter of fact does not necessarily lead to undersupply but rather pushes the knowledge creating firm to use it as an intermediary input for the sequential production of economic goods. Downstream vertical integration is the remedy to the problems raised by the nonappropriability and low tradability of knowledge as an economic good.[10] The generation of appropriate quantities of knowledge can be stimulated by the opportunities in the markets for the products that are manufactured and delivered by means of the technological knowledge they embody.

When technological knowledge can be easily appropriated by the innovator, either because of its complexity and hence natural levels of high appropriability, or because the regime of intellectual property rights is effective and easily enforced, knowledge transaction costs are low and, for given levels of internal coordination costs, firms prefer to directly sell the technological knowledge as a good per se in the markets for knowledge. Transaction costs in the markets for knowledge are lower than the costs of the internal coordination of the production of the product that embody that technological knowledge (Antonelli, 2004).

Knowledge fungibility is defined by the variety of production activities to which the same unit of knowledge can be successfully applied. With given knowledge transaction costs firms, able to introduce technological innovations with high levels of fungibility, are likely to be larger and more diversified and integrated. Strong increasing returns take place in the usage of the same stock of technological knowledge and can counterbalance the increase in average coordination and manufacturing costs.[11]

When technological knowledge is characterized by high levels of cumulability, so that the generation of each new unit of knowledge relies upon the localized accumulation of technological knowledge, dynamic coordination and transaction costs emerge. Dynamic transaction and coordination costs are defined in terms of opportunity costs of the governance of the stock of knowledge with respect to the stream of generation of new knowledge.

Inclusion now yields the opportunity to appropriate the eventual benefits stemming from the accumulation of knowledge in terms of higher opportunities for the introduction of additional units of knowledge. Exclusion and transaction instead yield new costs in terms of the missing opportunities to benefit from the cumulative learning processes associated with the production process itself. Firms select inclusion and exclusion not only with respect to the static assessment of coordination, transaction and production costs for a given product and a given technology, but also and mainly with respect to the technological opportunities that are associated with the learning processes (Antonelli, 2003a).

Knowledge transaction costs also matter on the demand side. Important resources can become necessary in order to search for, identify and purchase the bits of external knowledge that are necessary for the generation of new knowledge. Knowledge transaction costs are especially relevant when technological knowledge is characterized by high levels of complexity: each new bit of knowledge is the result of the recombination of many different elements. Knowledge transaction costs here affect the choice between making all the diverse bits of knowledge or purchasing them in the markets for technological knowledge. Intellectual property rights can perform the essential informational role of signalling, spreading the information that the knowledge corresponding to a patent exists and can be acquired.[12]

Knowledge stickiness is found when it is difficult to separate the knowledge, often tacit, from the human capital and the organizational routines of the unit where learning activities have been taking place and the knowledge has been generated. In this case an issue of indivisibility emerges. Financial markets and more generally the markets for property rights provide an opportunity for a firm that cannot directly exploit the new knowledge because of steep organization costs curves. The incorporation of the unit into a new corporation and its sale in the financial market becomes a viable solution. Here technological knowledge is embodied in the corporate structure.

This analysis leads to yet another dynamic aspect of the model considered concerning the economics of knowledge spillover. Firms cannot include the full range of activities engendered by their learning processes in manufacturing because of the limitations of organizational factors. A selection process takes place. The decision of inclusion takes into account both the profitability of the incremental activity and its organizational costs. With a positive slope of unit organizational costs, the inclusion of new activities can be rejected because of their high marginal organization costs, even if their profitability is above average levels. This paves the way to a new approach to knowledge spillover.

Knowledge spills from firms not only because of low appropriability, but also because of high internal selection standards, imposed by organizational costs. The larger the firms, the larger the spillover is likely to be: large firms are likely to have higher levels of coordination costs. Spillovers are likely to be larger the lower the market tradability of the knowledge. When knowledge tradability is higher, in fact, firms will be able to try and sell the marginal knowledge in the marketplace. When knowledge tradability is low and yet inclusion cannot take place because of coordination costs, firms are not able to take advantage of such technological opportunities. Such technological knowledge is then likely to 'spill' in the atmosphere and other firms, especially spin-offs, can take advantage of it.

The analysis of the interdependence between the laws of accumulation of competence and knowledge, their effects of the production process and the organization of the bundle of activities retained within the borders of the firm, makes it clear that at each point in time an equilibrium point between contrasting forces can be identified. Yet understanding the dynamics of the learning processes, which constitute the essence of the firm, and their effects in terms of the introduction of competence, knowledge and innovations, makes it clear that an equilibrium point is nothing more than a step into a path of continual transformation.

All differences in the localized rates of learning, accumulation of knowledge and competence, across the different modules and activities retained within the firm, and with respect to other agents in the marketplace, are likely to change its borders and the architecture of the organization. At the same time it is now clear how the rates of accumulation of localized knowledge in coordination and transaction have a direct bearing on the actual possibility of each firm to benefit from the accumulation of technological knowledge and to generate successful technological innovations. From this viewpoint, technological change is localized by the interplay between dynamics of technological learning and the dynamics of organizational learning.

5. CONCLUSION

Transaction costs economics and the resource-based theory have contributed along parallel lines of enquiry into the nature of the firm. In resource-based theory, the firm is viewed as a bundle of activities defined by their complementarity with respect to the generation of new knowledge and competence. In transaction costs economics, the firm is also a bundle of activities defined by given and exogenous costs of coordination and transaction. The merging of these research programs into a broader economics of governance is fruitful from many viewpoints.

The integration of the dynamics of accumulation of localized technological knowledge and the dynamics of the introduction of technological and organizational innovations is a necessary step towards a more articulated theory of the firm. The essential understanding of the basic trade-off between inclusion and exclusion, elaborated along the lines set forth by Ronald Coase and developed systematically by Oliver Williamson, can be further implemented in a more dynamic context.

The approach to the firm as a bundle of interdependent activities, where the generation of knowledge, production, coordination and transaction are complementary aspects of a broader process of governance, can be developed into a dynamic framework where the firm is viewed as a bundle of activities characterized by localized learning. Such an approach yields useful outcomes in terms of the systemic understanding of the interdependence and reciprocal feedback between different and yet complementary aspects of decisionmaking within the firm.

The economics of governance approach makes it possible to integrate the effects of the internal attributes of the firm in terms of the generation of coordination and transaction competence with the understanding of the external conditions of the markets both from a competitive and an informational viewpoint. Finally the understanding of the characteristics of technological knowledge and its generation process in terms of levels of appropriability, tradability, fungibility and complexity can be operationalized and integrated into a broader context.

The integration of transaction costs economics with the resource-based theory of the firm seems able to appreciate the variety of constraints and incentives, provided by the complexity of the organization and the marketplace, which shapes the working of the firm, viewed as the basic engine for the generation of knowledge immediately relevant for economic action in a market economy.

The economics of governance makes it possible to better understand the role of localized knowledge in the activities of coordination, transaction and production. In so doing it marks progress with respect to transaction costs economics, where both technological and organizational knowledge are exogenous and given. The governance economics approach however, makes it possible to better grasp the effects of the interactions between organizational and technological knowledge and the constraints raised by organizational factors such as coordination and transaction costs in shaping the process of accumulation and generation of new knowledge. In so doing the economics of governance makes possible a step forward with respect to the resource-based theory of the firm.

NOTES

1. See the useful surveys of Zingales (2000) and Mueller (2003) which, interestingly, do not take into account the role of learning in the theory of the firm and more generally ignore the resource-based theory of the firm.
2. Much attention has been paid in institutional economics to assessing the relationship between transaction costs economics and the incomplete contract theory (Grossman and Hart, 1986; Hart, 1995). Incomplete contract theory stresses the role of bounded rationality and the limitations of information impactedness in designing 'perfect' contracts and hence the need for internal coordination. Repeated renegotiations however can reduce the costs of the use of the markets. Upon these bases incomplete contract theory seems to be complementary to transaction costs economics rather than a substitute (Brousseau and Glachant, 2002).
3. The closer the market prices to the internal costs of manufacturing the more relevant is the ratio of transaction costs to coordination costs in defining the borders of the firm and the size of the portfolio of activities which it is profitable to include within its borders. When the market prices differ from internal manufacturing costs there is a direct incentive to change the borders of the firm. Such a difference may depend upon a variety of factors. Two classes can be easily identified: external factors and internal ones. The former

concerns the conditions of the marketplace; the latter the internal conditions of the manufacturing process. Imperfect market conditions in the supply of inputs and hence market prices that differ from the minimum average costs levels push either towards inclusion – when market prices are above minimum average costs – or towards exclusion when market prices for complementary products, often in the case of barriers to exit are below the minimum average costs levels. Internal factors in manufacturing matter as well in assessing the choice whether to include or exclude. Internal manufacturing costs may differ from market prices due to a variety of factors that belong to the idiosyncratic characteristics of the firm. Idiosyncratic increasing returns can explain internal manufacturing costs that are lower than market prices. This is clearly the case when factors of indivisibility and irreversibility, specific to the history of each firm, lead to economies of scale, economies of density, economies of scope and agglomeration externalities. When increasing returns apply to a product, the firm has a powerful incentive to include its production process. Clearly with decreasing returns in the manufacturing of a specific product the firm has a strong incentive towards its exclusion.

4. In this approach technological change is the endogenous outcome of the creative reaction, to the mismatch between expectations and actual facts, of myopic firms that are not bounded to quantity–price adjustments, but are able to change their technology also in a limited technical space defined by the pervasive role of irreversibility of fixed production factors and the effects of bounded rationality and learning processes. As a consequence at each point in time the marketplace is kept in disequilibrium between one possible equilibrium and many alternative ones introduced in a continual variety of efforts and attempts by heterogeneous and creative agents surprised by the mismatch between expectations and actual product and factor markets. The introduction of technological changes is an endless process because each innovation modifies the context anticipated by each other agent and hence induces other innovations. The process is path-dependent because at each point in time irreversibility constrains the decisionmaking of actors and yet their creative reaction can engender solutions that cannot be fully anticipated from their past. The assumptions about the irreversibility, of at least some inputs, and the key role of learning qualify the process as nonergodic: historic time matters. The assumptions about failure-induced technological change based upon reactivity, creativity and endogenous innovative capability mark the distinction between a past-dependent process and a path-dependent one: each innovation cannot be fully predicted from the past of the innovator.

5. This model builds upon the basic intuition provided by Riordan and Wiliamson (1985), subsequently elaborated in Antonelli (1999b).

6. This makes it possible to integrate the role of organizational knowledge with technological knowledge (Turvani, 2001).

7. In this context, the quality of markets in terms of trust, information and public knowledge is an essential component of the endowment of social capital (Dasgupta and Seragekdin, 2000). The quality of the markets varies according to their thickness: the number of players on both the demand and the supply sides. Industrial dynamics, in terms of rates of entry and exit, may impose additional burdens in terms of transaction costs even if it has positive effects in terms of the reduction of the market prices towards competitive levels. The better the quality of the markets, from an informational viewpoint, the lower is the amount of search costs to identify the correct price and reliable partners in trade. The levels of opportunism socially accepted and hence the levels of trust that are necessary to stay in the marketplace are clearly relevant. Institutions and norms hence enter the scene. The complexity and the novelty of the products and hence the amount of information that are necessary to assess their quality play a major role in the definition of transaction costs. Institutional and social conditions play a key role also in assessing the levels of coordination costs (Antonelli, 2003a).

8. All changes in the conditions in which transactions, interactions and coordination take place can be considered as changes in the endowment of social capital and in its structure. When the thickness of the markets increases as well as their informational transparency and the levels of trust, transaction costs decline and hence the use of the markets become more effective: the borders of the firm shrink. From a comparative viewpoint the

differences, across regions and industries, in the organization and in the size of the firms can be now viewed as determined by the differences in the thickness, transparency and competitiveness of the markets. Countries and industries where the average size of firms is larger as well as the scope of their portfolio are likely to be characterized by lower levels of transparency. The size of firms is smaller as well as the levels of diversification in industrial districts typically characterized by high levels of trust and transparency, mainly because of a historic tradition of repeated interactions. Large diversified holding companies can be considered the end result of lower levels of informational quality of the markets for intermediary inputs. Firms may grow into large diversified companies however, also when coordination competence is very high and it is rooted in the national and industrial traditions and institutions: as such it is difficult for it to swarm elsewhere (Nooteboom, 2002).

9. See Antonelli et al. (2003) for an analysis of the international markets for technology and an empirical estimate of the role of external knowledge.
10. This result is important as it is in contrast with the traditional argument about the failure of markets, as a coordination system, in the allocation of resources to the production of knowledge because of the lack of incentives stemming from low appropriability and the related 'knowledge as a public good' tradition of analysis (Antonelli, 2004).
11. The welfare losses stemming from high knowledge transaction costs and hence high levels of vertical integration in the case of high levels of knowledge fungibility are high because the application of each bit of fungible knowledge to other activities is limited by the embodiment in a firm active in a narrow range of products.
12. The design of an intellectual property regime that makes possible the application of the liability rule and hence the reduction of the exclusivity of ownership combined with the full valorization of the informational role of patents can help in reducing knowledge transaction costs (Antonelli, 2004).

REFERENCES

Alchian, A. A. and H. Demsetz (1972), 'Production information costs and economic organization', *American Economic Review*, **62** (5), 777–95.

Antonelli, C. (ed.) (1988), *New Information Technology and Industrial Dynamics*, Boston, MA and Dordrecht: Kluwer Academic.

Antonelli, C. (1999a), *The Microdynamics of Technological Change*, London: Routledge.

Antonelli, C. (1999b), 'The organization of production', *Metroeconomica*, **50** (2), 234–53.

Antonelli, C. (2001), *The Microeconomics of Technological Systems*, Oxford: Oxford University Press.

Antonelli, C. (2003a), *Economics of Innovation: New Technologies and Structural Change*, London: Routledge.

Antonelli, C. (2004), 'The governance of localized technological knowledge and the evolution of intellectual property rights', in E. Colombatto (ed.), *The Elgar Companion to Property Rights Economics*, Cheltenham, UK and Northampton, MA: Edward Elgar.

Antonelli, C., R. Marchionatti and S. Usai (2003), 'Productivity and external knowledge: the Italian case', *Rivista Internazionale di Scienze Economiche e Commerciali*, **50**, 69–90.

Arora, A., A. Fosfuri and A. Gambardella (2001), *Markets for Technology*, Cambridge, MA: MIT Press.

Argyres, N. S. (1995), 'Technology strategy, governance structure and interdivisional coordination', *Journal of Economic Behavior and Organization*, **28** (3), 337–58.

Arrow, K. J. (1962a), 'The economic implications of learning by doing', *Review of Economic Studies*, **29**, 155–73.

Arrow, K. J. (1962b), 'Economic welfare and the allocation of resources for invention', in Richard R. Nelson (ed.) *The Rate and Direction of Inventive Activity: Economic and Social Factors*, Princeton, NJ: Princeton University Press for NBER, pp. 609–25.

Arrow, K. J. (1969), 'Classificatory notes on the production and transmission of technical knowledge', *American Economic Review*, **59** (2), 29–35.

Arrow, K. J. (1974), *The Limits of Organization*, New York: W. W. Norton.

Baldwin, C. Y. and K. B. Clark (2000), *Design Rules: The Power of Modularity*, Cambridge, MA: MIT Press.

Bonazzi, G. and C. Antonelli (2003), 'To make or to sell? The case of in-house outsourcing at FIAT Auto', *Organization Studies*, **24** (4), 575–94.

Brousseau, E. and J. M. Glachant (eds) (2002), *The Economics of Contracts*, Cambridge: Cambridge University Press.

Chandler, A. D. (1962), *Strategy and Structure: Chapters in the History of the Industrial Enterprise*, Cambridge, MA: MIT Press.

Chandler, A. D. (1977), *The Visible Hand: The Managerial Revolution in American Business*, Cambridge, MA: Belknap Press.

Chandler, A. D. (1990), *Scale and Scope: The Dynamics of Industrial Capitalism*, Cambridge, MA: Belknap Press.

Chandler, A. D., P. Hagstrom and O. Solvell (eds) (1998), *The Dynamic Firm: The Role of Technology Strategy Organization and Regions*, Oxford: Oxford University Press.

Coase, R. H. (1937), 'The nature of the firm', *Economica*, **4**, 386–405.

Dasgupta, P. and I. Serageldin (eds) (2000), *Social Capital: A Multifaceted Perspective*, Washington, DC: The World Bank.

Foss, N. J. (1997), *Resources Firms and Strategies: A Reader in the Resource-based Perspective*, Oxford: Oxford University Press.

Foss, N. J. (1998), 'The resource-base perspective: an assessment and diagnosis of problems', *Scandinavian Journal of Management*, **15** (1), 1–15.

Foss, N., and V. Mahnke (eds) (2000), *Competence Governance and Entrepreneurship. Advances in Economic Strategy Research*, Oxford: Oxford University Press.

Grossman, S. J. and O. D. Hart (1986), 'The costs and benefits of ownership: a theory of vertical integration', *Journal of Political Economy*, **94** (4), 691–719.

Guilhon, B. (ed.) (2001), *Technology and Markets for Knowledge: Knowledge Creation, Diffusion and Exchange within a Growing Economy*, Boston, MA and Dordrecht: Kluwer Academic.

Hart, O. D. (1995), *Firms, Contracts and Financial Structure*, Oxford: Oxford University Press.

Holmstrom, B. (1989), 'Agency costs and innovation', *Journal of Economic Behavior and Organization*, **12**, 305–27.

Lamberton, D. (ed.) (1971), *Economics of Information and Knowledge*, Harmondsworth: Penguin.

Lazonick, W. (1991), *Business Organization and the Myth of the Market Economy*, New York: Cambridge University Press.

Loasby, B. J. (1999), *Knowledge Institutions and Evolution in Economics*, London: Routledge.

Loasby, B. J. (2002), 'The division and coordination of knowledge', in Sheila C. Dow and J. Hillard (eds), *Post Keynesian Econometrics, Microeconomics and the Theory of the Firm, Beyond Keynes*, vol 1, Cheltenham, UK and Northampton, MA: Edward Elgar, pp. 6–15.

Menard, C. (ed.) (2000), *Institutions Contracts and Organizations. Perspectives from New Institutional Economics*, Cheltenham, UK and Northampton, MA: Edward Elgar.

Mueller, D. C. (2003), *The Corporation. Investment, Mergers and Growth*, London: Routledge.

Nooteboom, B. (2000), *Learning and Innovation in Organizations and Economics*, Oxford: Oxford University Press.

Nooteboom, B. (2002), *Trust: Forms, Foundations, Functions, Failures and Figures*, Cheltenham, UK and Northampton, MA: Edward Elgar.

Penrose, E. T. (1959), *The Theory of the Growth of the Firm*, Oxford: Basil Blackwell.

Riordan, M. and O. E. Williamson (1985), 'Asset specificity and economic organization', *International Journal of Industrial Organization*, **3**, 365–78.

Simon, H. A. (1947), *Administrative Behavior*, New York: Free Press.

Simon, H. A. (1962), 'The architecture of complexity', *Proceedings of the American Philosophical Society*, **106**, 467–82.

Simon, H. A. (1969), *The Sciences of the Artificial*, Cambridge, MA: MIT Press.

Simon, H. A. (1982), *Metaphors of Bounded Rationality. Behavioral Economics and Business Organization*, Cambridge, MA: MIT Press.

Spulber, D. (1999), *Market Microstructures: Intermediaries and the Theory of the Firm*, Cambridge: Cambridge University Press.

Teece, D. J. (1986), 'Profiting from technological innovation: implications for integration, collaboration, licensing and public policy', *Research Policy*, **15**, 285–305.

Teece, D. J. (1998), 'Capturing value from knowledge assets: the new economy, markets for know-how and intangible assets', *California Management Review*, **40** (3), 55–79.

Teece, D. J. (2000), *Managing Intellectual Capital*, Oxford: Oxford University Press.

Turvani, M. (2001), 'Microfoundations of knowledge dynamics within the firm', *Industry and Innovation*, **8** (3), 309–23.

Williamson, O. E. (1975), *Markets and Hierarchies: Analysis and Antitrust Implications*, New York: Free Press.

Williamson, O. E. (1985), *The Economic Institutions of Capitalism: Firms, Markets, Relational Contracting*, New York: Free Press.

Williamson, O. E. (1990), 'The firm as a nexus of treaties: an introduction', in M. Aoki, B. Gustaffson and O. E. Williamson (eds), *The Firm as a Nexus of Treaties*, London: Sage, pp. 1–25.

Williamson, O. E. (1996), *The Mechanisms of Governance*, New York: Oxford University Press.

Zingales, L. (2000), 'In search of new foundations', *Journal of Finance*, **55** (4), 1623–53.

3. Innovation, consumption and knowledge: services and encapsulation

Jeremy Howells[1]

1. INTRODUCTION

The role that consumption plays in the innovation process within the firm and in the formation remains a neglected aspect of a firm's capabilities. Thus consumption and the way firms consume intermediate goods and services form an important, but neglected, part of a firm's capability set. The role of services is highlighted in this review and discussion. This is because by introducing a service dimension to the discussion about innovation, a new perspective is shed on the process of consumption and its relationship within innovation within the firm. This is for three interrelated reasons. Firstly, it is suggested that services are important in the consumption of new goods (and services). Secondly, the way (i.e. the routines that they can potentially develop) firms consume (intermediate) goods yields service-like attributes and these form important and distinctive capabilities for the firm. Lastly, related to this, the process of consumption and the development of routines associated with this process are forms of disembodied, service innovations.

The analysis seeks to focus on the role of consumption in influencing innovation in intermediate goods and services. It is presented here that this is a neglected field of research for a number of reasons. Firstly, despite a number of studies, the role of consumption and demand in the innovation process still remains largely neglected. Secondly, and in particular, the role of consumption in service innovation has only been briefly commented upon. This is particularly true in connection with the consumption by firms of intermediate services and goods, where most of the discussion on service consumption has been in respect of final consumption by individuals and households (see, for example, Gershuny, 1978; Gershuny and Miles, 1983).[2]

This chapter therefore seeks to put forward an exploratory framework in relation to the role that consumption may play in services in the innovation process within the firm exploring existing literature and using primary case study material. In this context, the analysis highlights the important role that services play in the consumption of new goods by firms, and the interplay and blurring between goods and services in consumption (although this is not a new phenomenon as will be shown). The focus, therefore, as will be explained, is primarily on the consumption of intermediate goods and services by firms, and not in the role of final consumption in innovation per se. To this end this chapter is structured as follows. Section 2 discusses the nature and role of consumption in services. Section 3 expands the analysis to consider the relationship between consumption and services and innovation. Section 4 examines what elements of the consumption process in relation to innovation have been studied already. Section 5 concludes.

2.   CONSUMPTION AND SERVICE CONSUMPTION

From a period of relative stasis in terms of academic interest and progress, there have been a number of recent analyses which have sought to review and critique existing consumption theory particularly from a neoclassical perspective but also more generally from other reformulated economic studies. These studies have sought to further develop and integrate consumption theory within a wider socioeconomic context, in particular focusing on consumption as a change agent.

It is not the intention here to further review these studies in detail. However, these more recent analyses have highlighted a number of characteristics of consumption building upon the existing body of literature that have not been readily acknowledged in the past. In brief, these are that consumption is an active, rather than passive process, with consumers actively seeking novelty to satisfy needs and tastes (Bianchi, 1998, p. 65 and pp. 75–81). Consumers therefore act as interactive agents in the wider competitive environment (Gualerzi, 1998, p. 59). However, effective consumption patterns require time and resources to develop (Loasby, 1998, p. 94) and this sets constraints on such trial-and-error learning (Metcalfe, 2001, p. 44). In this way consumption activity can also be seen as forming a key capability of the firm (Langlois and Cosgel, 1998, pp. 110–11), which involves a process of learning for consumers (Loasby, 1998, p. 98; Robertson and Yu, 2001, p. 190; Witt, 2001, pp. 28–31). It can also be seen in the development of efficient routines (Langlois and Cosgel, 1998, p. 59; Langlois, 2001, p. 90) for successful consumption.

These above points highlight a sea change in thinking about the nature, value and importance of consumption in shaping economic change and including innovation. In particular, such studies stress that to be an effective consumer involves time and resources. Thus they partially take-up Lancaster's (1966, 1971, 1991) theory of consumer choice, in that the consumer needs to be an active agent and invest in time and resources to build up capabilities to consume effectively. These studies, in highlighting the development of consumer competencies and routines, also echo and build upon Stigler and Becker's (1977, p. 78) neoclassical concept of the accumulation of 'consumption capital'. Just as innovations require considerable investment to produce, so do consumers need to invest in new capabilities and routines to consume them.

Unfortunately such studies, although extremely significant in the conceptualization of consumption (particularly from a wider economic viewpoint), have two major drawbacks. They, firstly, only tangentially discuss the role of consumption in the process of innovation. The issue of innovation and technology, if noted at all, is within the broader context of the search for new tastes or in the desire for novelty. The search for novelty and the development of new tastes are important issues which have been neglected in orthodox, neoclassical economics, but their impact on innovation is not pursued. Secondly, such studies make little or no specific reference to the consumption of services or to intermediate goods, but instead remained focused on final consumption patterns by the individual or households. Thus, the literature seems to stick to the issues of the 'atomistic' individual or (at best) household consumer and their consumption of final goods (see, for example, Earl, 1986, pp. 20–1). There are a few exceptions to this neglect of intermediate goods and services, most notably by Langlois and Cosgel (1998; Langlois, 2001) in their development of a 'capability model' of consumption associated with the development of consumption routines which refers both to services (Langlois and Cosgel, 1998, p. 109) and intermediate goods (pp. 89 and 114) but again, although excellent, they do not cover these issues in any detail (however, see below). [3]

Although there has been little direct focus on the consumption of services, there are a number of underlying assumptions in the literature associated with services and their consumption. These assumptions in particular cover three areas: firstly, the definition of services and their differentiation from goods; secondly, the nature of consumption itself associated with the notion of utility in what is consumed (and in turn involving the temporal dimension to consumption); and thirdly, linked to this, the co-joint nature and demand of goods and services (in contrast to the first perspective).

Firstly, how services are consumed, although not outwardly acknowledged, has been used as a dimension in helping to define what services are. One of the key distinguishing features of services (together with immateriality and perishability or nonstorability) has been the notion of the simultaneity of production and consumption with regard to services

(Petit, 1986, p. 9). Thus, Hill (1977, p. 337) notes: 'Services are consumed as they are produced in the sense that the change in the condition of the consumer unit must occur simultaneously with the production of that change by the producer: they are one and the same change. The consumption of a service cannot be detached from its production in the way that the acquisition of a good by a consumer in an exchange transaction may take place some time after the good is produced.' This simultaneity in consumption has been coupled with the requirement for the physical presence and co-location of the production and consumption (see, for example, Quinn and Dickson, 1995, p. 344). However, this view is not accepted by others who stress that the physical proximity of a provider and receiver is not a ubiquitous requirement (Petit, 1986, p. 9; Sampson and Snape, 1985, p. 172). More fundamentally here, Greenfield (1966, pp. 8–9) questions the issue of simultaneity of consumption and production and goes on to suggest that many services are not perisherable and yield utility over relatively long timespans. The issue, therefore, is useful in terms of helping to differentiate between some services and other forms of economic activity, however it is less helpful in continuing to suggest that the consumption (in terms of yielding utility) of services is an instantaneous process.

Secondly, concerning the more fundamental issue of what the nature of consumption is, it is useful to review the long established concept of utility that has been used by economists and which highlights the distinction between desires and the satisfaction of wants.[4] What discussion there has been about the consumption of services, has been in a very one dimensional and direct sense associated with the 'act of buying' a specific service product. However, as economists have long recognized in relation to goods, these goods satisfy certain wants and therefore yield a utility in satisfying these wants. In this context many services may be seen as 'purer' in utilitarian terms as they are often nearer in the spectrum of satisfying ultimate wants, whereas goods may be seen as being more about an interim milestone on the road to such satisfaction (or more specifically provide a solution to a problem; Gadrey et al., 1995, p. 5). One may think of purchasing a television set which satisfies an initial desire, but which can only be satisfied (partially or completely depending on what television programme you watch!) by viewing television programmes on the television. This can lead into a deeper metaphysical discussion about what consumption fundamentally is. However, it also raises the notion that consumption has a strong temporal quality (and with it, dynamic and evolutionary qualities). Consumption is rarely instantaneous (i.e. when a good or service is actually purchased) in the sense of satisfying immediate wants (Robinson, 1962, pp. 122–3).

This temporal dimension to consumption is somewhat at odds with the definition used of services, namely, that production and consumption of services (or at least most of them) has to be simultaneous. If we interpret consumption here as that of the simple purchase of services this may hold true, but is more tenuous, or simply incorrect, if we hold that consumption is a much richer and time consuming process about satisfying more fundamental wants. This is difficult to illustrate adequately, but even such a seemingly ephemeral and transient matter as watching a film at the cinema is a case in point. Should its 'consumption' solely be seen as watching the film or does it encompass the much wider consumption experience of discussing the film with friends and colleagues afterwards and remembering it in connection with other films and books after the event? With the exception of Greenfield (1966, pp. 8–9) few researchers in the services field have recognized the temporal and durable notions of the consumption process.

However, this leads into a discussion about the similarity and interlinked nature of goods and services (rather than their distinctiveness raised earlier). This can be seen on two levels. Firstly, if one is interested in consumption as satisfying ultimate wants, goods actually can have, or fulfil, service-like attributes. Thus, Lancaster (1966, p. 133) notes: 'The chief technical novelty lies in breaking away from the traditional approach that goods are the direct objects of utility and, instead, supposing that it is the properties or characteristics of the goods from which utility is derived. We assume that consumption is an activity in which goods, singly or in combination, are inputs and in which the output is a collection of characteristics.' This issue is developed by Saviotti and Metcalfe's (1984) work on attempting to measure technical change of products, i.e. material artefacts. Saviotti and Metcalfe (1984), building upon Lancaster's (1966) work, stress that a product can have both internal properties, i.e. those relating to the internal structure of the product, and external properties, relating to wider issues associated with the type of service being offered to users as part of that good. Thus, goods have 'service' characteristics associated with them (see, for example, Bressand, 1986; Bressand and Nicolaidis, 1989; Hill, 1977; Silvestro et al., 1992).

If goods are seen as having service qualities in their consumption and utility, it has also been recognized that products and services have long been associated together, not just to support goods (De Brentani, 1995), but more fundamentally they are interlinked more directly with both in their co-production (Bettencourt et al., 2002) and co-consumption (Howells, 2001). Indeed, this takes up Hill's (1977) notion of services as changing the condition of the consumer. Thus, as far back as 1892 Alfred Marshall highlighted the issue of derived demand and joint demand for goods and services (Marshall, 1899, pp. 218–23), exemplified in his notion of composite demand. More specifically this has been explored in more detail by Swann (1999) in his analysis of 'Marshall's consumer' and the increasing levels of sophistication that consumers can present in the consumption process. This has also been acknowledged in relation to the growth and nature of inter-mediate or producer services, which has been most developed within the geographical literature (see, for example, important contributions by Britton, 1990; Daniels and Bryson, 2002; Wood, 1991, 1996).

3.   CONSUMPTION, SERVICES AND INNOVATION

The above has attempted to highlight the nature and role of consumption in services (and indeed services in consumption). This may now be extended to consider a tripartite relationship involved with the role of consumption as it relates to services and innovation. This will be viewed first within the context of how services influence the consumption of innovation, before discussing the influence of consumption on service innovation.

On the basis of the above discussion about services and consumption, it is presented here that services can play a key intermediary and conduit role in the innovation process in respect of both new goods and services. In short, services often encapsulate, or act as 'wrappers' to, goods and resources. This section will firstly briefly explore the basic principle of encapsulation (see Howells, 2000, 2001, 2004 for more detailed reviews) before moving more specifically to analysing its role in relation to the innovation process.

Part of the process of 'servicization' is the trend in manufacturing firms (and indeed in agricultural and resource-based companies) towards provid-ing service products that are related to the manufactured products they produce. Vehicle manufacturers, for example, have created finance and leasing subsidiaries to facilitate the purchase of their cars and trucks. They also have substantial maintenance and repair operations associated with aftersales care. Through their sales franchises and outlets, they may buy back cars and trucks for secondhand sales. More recently, owing to increased environmental awareness and legislation, vehicle producers arrange for the disposal and recycling of their cars (see Ford and Fiat examples in Table 3.1). All these activities are closely associated with selling the manufactured product, the car, but they also respond to consumers' wishes in terms of support (see Shostack, 1977, p. 74).

This trend is also evident in the aerospace industry, with aircraft builders offering finance and leasing arrangements. Aerospace engine manufac-turers not only provide finance but also operate major repair and overhaul facilities. In this industry, General Electric (GE), for example, has a major finance and leasing company (GE Capital Services) and also operates an assortment of purchasing, leasing and rental options.[5] More specifically, GE Engine Services (GEES) offers a wide ranging maintenance service package called 'GE On Wing Support' which provides engine inventory, long term preservation and facilities support. Equally, Rolls-Royce has moved strongly into acquiring aero engine repair and maintenance companies across the world, such as National Airmotive, a US engine repair company. Increasingly, aerospace engine manufacturers are providing engines not as a product (an engine) but as a service (hours of flight). This aspect of the 'servicization' phenomenon may be termed the 'service encapsulation' of goods and materials. As such, manufactured products are not offered to consumers in their own right but rather in terms of their wider service attributes.

This can occur in two ways. The first is to offer the manufactured product along with closely aligned service products in a single package. In the case of the motor car, this means finance, insurance, maintenance warranties, repurchase clauses and tax all rolled together. The second is more sophisticated in that it seeks to offer the consumer not the manufactured product itself but rather the goal that the purchase of the manufactured product would ultimately fulfil. A case in point is the replacement of aerospace engines (product) by hours of flight (service) by both General Electric and Rolls-Royce noted above. Another example, taken from the computer industry, is to offer computer services to carry out certain tasks rather than supply the computers that are used to provide the service. Zeneca (now AstraZeneca) bought Salick Health Care (SHC), a company that operates fully integrated cancer and chronic care services in the United States. By so doing, AstraZeneca can both better monitor the performance of its own cancer drugs and that of its competitors and also test the notion of offering more complete healthcare services to customers (patients). As part of this total patient care service, AstraZeneca also operates in the United States a managed care service, SalickNet, to provide customized disease management programmes.

These examples suggest that ultimate consumption is being satisfied at a different but more effective (higher) level of utility by going directly to the central issue of concern for both the intermediate actor (the healthcare service or insurance company) and the ultimate customer (patient). In addition, these companies offer something beyond what their competitors offer. General Electric not only provides a lifecycle solution for all a customer's healthcare equipment via GE Healthcare Services but has also moved to provide a remote diagnostic service, as well as medical diagnostic equipment (Table 3.1). Indeed, it has moved to provide a wider array of medical services associated with more general healthcare operations in management and decision support services.

Such encapsulation mechanisms are being combined to allow firms to provide consumers with ever more seamless solutions and to create more value added for the firm that supplies them. However, this goes beyond issues of industry outsourcing and vertical integration and also moves beyond the economic and competitive benefits of integrating the supply chain to focus on satisfying customers' actual demands. For firms that sell such goods and services, these activities suggest a new concept of what consumption and innovation are about. In the case of the car, consumption has moved from the simple, one-off purchase to the wider process of buying, using and maintaining a car over the long term. This shift in focus has major implications for firms that sell such products and services in terms of how they address consumers' needs and satisfy their ultimate demand.[6] Consumption is therefore not a one-off contact via the sale of a product but a continuing process involving long term customer contact through service delivery. This is to be expected if consumption shifts from a single, one-off act to long term user support; i.e. from selling a car at a single point in time to supplying fast/reliable/cheap/flexible/safe transport over a period of time.

Figure 3.1 provides a diagram of various services that may be sold with a manufactured product over its lifetime. It includes a variety of stages. Firstly, it can involve a sphere of services that are important prior to purchase in terms of setting up and facilitating purchase and delivery in a convenient and timely fashion. It may involve complex turnkey operations for fixed assets and sending out specialist technicians and advisers to show how the equipment and plant is run. This in turn can be associated with quite long term training programmes for operatives and a clearly designated handover period. Mathé and Shapiro (1993, p. 5) have indeed highlighted the crucial role of aftersales and delivery services in sustaining the long term success of manufacturing firms. Secondly, it can then involve services required in using the product (for example, its efficient operation) and this crucially centres around customer support services associated with the necessary support of the good or product whilst in use. One such service is the technical support of the product, which should ideally be performed by the team who were also responsible for it during the product definition phase. On this basis, design knowledge is readily available to address maintenance problems, whilst the other way, feedback of experience in use allows the design to be improved on a continuous basis (Ruffles, 2000, p. 10)

Thirdly, in terms of the ongoing use of the good or product, are the more specific service activities of repair and maintenance. Here service enhancements and innovations in routines and practice can bring about major improvements to the performance and use of the product and hence attractiveness to the customer. One important area here has been in the emergence and development of predictive support services. The final support service that can be provided is to dispose of a product. This has become increasingly important for many goods, such as tyres, because of growing environmental concerns associated with waste disposal. Indeed aftersales and delivery services have been increasingly recognized as crucial to the long term success of manufacturing firms. This has been increasingly coupled with End-of-Life (EOL) disposal issues (Toffel, 2003). Thus GE Aircraft Engines operates an 'Exchange Engine' programme for its aircraft engines, where a customer can exchange an unserviceable or faulty engine for a serviceable engine, with the value of the unusable engine being credited towards the price of the new engine.

The process of encapsulation, associated with combinatorial aspects of goods and services can be seen to have two effects on innovation. This first is directly impacting the innovation process and how innovation is perceived and conducted by firms. Thus, the process of encapsulation can help radically change a company's perception of innovation. When companies move from simply selling a product to long term involvement in satisfying customers' needs, they will start to be more concerned, for example, about reliability and ease of servicing, the costs of which they may increasingly have to bear. For aerospace engine firms, reliability and safety have always been high on the agenda. However, contracts with consumers, such as Rolls-Royce's contract with American Airlines, where heavy penalties are incurred for unintended periods of inactivity due to engine problems, have meant that in-flight monitoring and diagnostics of various critical parts of an engine become more important. Thus, problems associated with particular components or performance can be relayed in flight and the maintenance teams and components can be ready when the plane arrives and problems can be solved before they create a major difficulty for the company and its customer. For companies like Rolls-Royce, therefore, instrumentation and electronics for monitoring and diagnostics become more central to the company's innovation profile and strategy. Its competitive and technological profile will place more emphasis on in-flight monitoring and fault finding, improved reliability, better organization of parts and components logistics and improved and faster maintenance. This involves what it terms 'predictive support' which combines the use of enhanced reliability prediction models with data feedback from actual service operations. A comprehensive predictive support service will also combine details of product and spares availability with inventory control, service scheduling and maintenance forecasting. Similarly, GE Aircraft Engines has spent a lot of resources on developing its Remote Diagnostics service which remotely monitors aircraft and engine information which can often identify issues before they become operational problems. This has meant building up development expertise in data monitoring and system integration and management techniques. The competitive focus has shifted towards developing and exploiting service qualities and moving into

Table 3.1 Service encapsulation of manufactured or resource-based products

Company	Manufactured product	Service encapsulator	Final offering and consumption
AstraZeneca	Cancer drugs	Cancer healthcare (Salick Health Care)	Cancer treatment
Fiat	Cars	Financial and insurance services for car customers (Toro Assicurazioni)	Car travel
Ford	Cars	Car travel services: financing and leasing (Ford Finance); maintenance (Kwikfit); in-vehicle services (Wingcast); web retailing (Fordjourney.com)	Car travel
General Electric	Aerospace engines	Leasing or selling hours of flight	Air travel
General Electric	Medical diagnostic equipment	Medical analysis and diagnosis	Diagnostic and medical services
Liebherr	Cranes	Special software programming to control and run the machines; remote running and testing	Lifting
Pacific Power International/ Rio Tinto Energy Resources	Coal	Power plant design and operating expertise and environmental advice	Power/Energy
Rolls-Royce	Aerospace engines	Leasing or selling hours of flight (minus time on ground due to faults)	Air travel
Xerox	Reprographic equipment	Maintenance and leasing	Photocopying

Source: Howells (2004)

nontechnological innovations. Intrinsic to this shift in the company's innovation strategy has been a de facto transition from a manufacturing to a service type of contract with the customer in the field of aero engines.

Secondly, encapsulation can more generically be seen to work on innovation via the process of consumption and this can happen in two ways. Firstly, through existing services encapsulating new goods (or services); and, secondly, through new services encapsulating old goods (or services). In relation to existing, familiar services encapsulating new goods (or services), these in turn can be seen as yielding a number of different effects. The first is what might be termed as the 'familiariser' effect where providing a familiar, trusted service to a new good, makes it more acceptable for consumers to adopt the innovation. The second property associated with this combinatorial process is the 'buffer' effect where once a good has been adopted it enables the consumption of a new good in exactly the same consumption service format in which the former, old good (i.e. earlier vintage) was consumed. The last type of change property here is the 'facilitator' effect which is associated with encouraging and helping consumers to learn new practices and routines through an existing service 'window' or framework to use a new good. As noted earlier, learning plays an important role in the consumption of new innovations.

However, new services, in turn, can encapsulate existing goods (or services). Here services are used to improve the acceptability, flexibility and performance of existing goods and these attributes are outlined in more detail below, using some simple examples. As such, new services encapsulating existing goods can provide a number of revitalizing and innovative roles. The first such change relationships is the 'sweetener' effect. Thus, by improving the acceptability of a good through a new service format it helps overcome obstacles which may have prevented the adoption and use of a good or service before. This may involve better set-up and operational instructions, which to the consumer may involve simple changes, but to the provider may involve complex, disembodied technical changes to routines and practices in the presentation of the good. Thus, technical documentation, including product configuration data, maintenance and operating instructions for use, may involve changes in highly complex organizational and operating routines both for the service being sold and the purchasing company using such new documentation. The second such property is the 'flexibility' effect. Here combining new services with an existing good (or service) may improve flexibility of use. Better maintenance practices and fault diagnostics may allow the good to be made available to the user over longer periods of time or during periods when it was previously not possible to use it.

The third such change property is the 'performance' effect, whereby a new service may improve the performance of the good. The most obvious

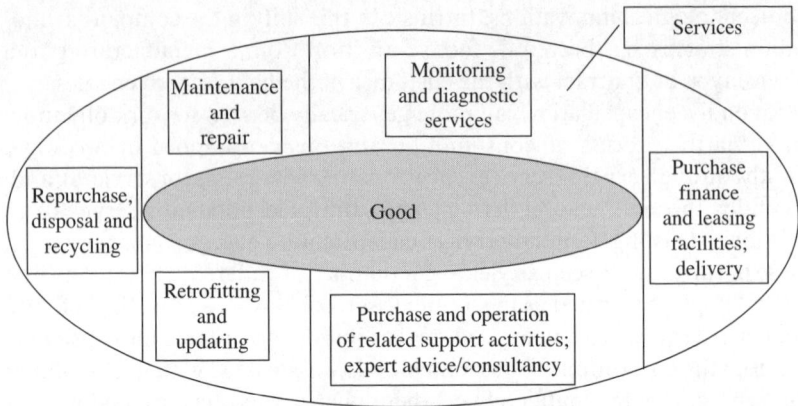

Source: Howells (2004)

*Figure 3.1   Service encapsulation: new patterns of consumption and
innovation through the life cycle of a good*

example here is a new software program (with improved performance and
functionality) being loaded to run on existing information technology
equipment. However, another example is the development of a whole new
area of services associated with 'predictive support services' which both
improve the efficiency of a good, but also reduce 'downtime' in its use. The
last effect is the 'functionality' effect. This change property or effect is where
a new service may allow an exsiting good to be used in a different way
(see, for example, Robertson and Yu, 2001, p. 188). Thus, a piece of testing
equipment, or scientific instrument (von Hippel, 1976), may be used to test
for things in different environments or conditions it was previously not initially designed for. However, this in turn may involve modifications to the
existing good, generating a new round of innovation.

Services in this way can be seen as playing an important role in the
consumption of innovations by enabling consumers to interact and accom-
modate these new goods and services more easily. Through mechanisms
like branding they can reassure consumers and act as pointers to quality
standards which were experienced through previous rounds of consumption. However, services also provide conduits through which innovations
are adopted and learning mechanisms through which innovations are used.
Adapting Scitovsky's (1992, p. 225) notion of skilled consumption here
adopting a new good requires some of its attributes to be recognized and
understood (and in this context, facilitated by services) if the consumption
of the innovation is to be successful.

Note too, there is often a complex interplay between services and goods in the innovation process; changes in one sphere often 'sparking off' changes in the other sphere. This can lead to constant rounds of new interactions (see von Hippel, 1994, pp. 432–4), with the innovation process between services and goods creating its own innovative dynamic. Indeed, staff from the user organization may form part of the production function helping to produce the services (O'Farrell and Moffat, 1991; see also O'Farrell, 1995). Strong behavioural forces also come into play here in terms of how consumers interact with products and how these change with the introduction of something radically new (Cooper, 2000, p. 4).

4. CONSUMPTION AND INNOVATION: THE DEVELOPMENT OF FIRM CAPABILITIES

What elements of the consumption process in relation to innovation have actually been studied? Those studies that could be said to cover aspects of the consumption process include buying or purchasing goods and services; their adoption, diffusion and absorption; and their use. More specifically:

1. There have been a whole series of studies that have highlighted the important function of purchasing in new product development and innovation (see, for example, Burt and Soukup, 1985; Thomas, 1994; Wynstra et al., 2000). This is centred on the initial act of buying and how technical inputs can be inputted into new product developments.
2. There has also been a growing number of studies more generally associated with adoption, diffusion and absorption (Cohen and Levinthal, 1990). These studies have tended to focus on decisionmaking processes, barriers to adoption and absorption capacities of the firm. More recently studies have sought to map out the diffusion of innovations within the firm after the initial purchase and adoption and reveal that this is far from an instantaneous or homogenous process (Pae et al., 2002).
3. The role of users in the innovation process has also been extensively studied (Foxall, 1987; Holt, 1987; Lundvall, 1992; Parkinson, 1982; Shaw, 1985, 1987; von Hippel, 1976, 1988). Here the important role that users play and interact with producers has been highlighted. However the focus has mainly been on the impact of users on producers in terms of what innovations they produce rather than so much on the actual use and shaping of innovations by consumers (however see Foxall, 1988).

4.   The improved consumption of goods and services in new or different ways to yield benefits to its operational efficiency has been addressed. This can be related to a whole literature on productivity and improving yields, particularly in relation to materials use.

However, there have been two further aspects of the consumption and innovation within the firm that have been largely overlooked. Thus there has been little or no discussion in relation to:

1.   The improved consumption, through new procedures and routines, of goods and services to improve their long term strategic position;
2.   Lastly, but significantly, an integrated and holistic view of consumption within the firm as it relates to innovation. Although the above has listed some elements of the consumption process which have been analysed there has been very little, if any, analysis 'in the round'.

In terms of these two latter points there has been little work. However, Edith Penrose has highlighted the role of consumption, in her book *The Theory of the Growth of the Firm*, in terms of defining a firm's competitive advantage and distinctiveness (and coincidentally the service-like attributes of goods and resources). Thus, Penrose (1995, pp. 78–9) notes:

> Physically describable resources are purchased in the market for their known services; but as soon as they become part of a firm the range of services they are capable of yielding starts to change. The services that resources will yield depend on the capacities of the men using them, but the development of the capacities of men is partly shaped by the resources men deal with. The two together create the special productive opportunity of a particular firm.

As such the distinctive way that firms consume and use physical goods and resources in terms of their service utility forms a key, but neglected element within a firm's repertoire of capabilities (see Richardson, 1972). Indeed how firms translate goods and resources into services may form an important component and attribute in this whole process. The process of consumption as a capability can form a powerful complementary asset in innovation that consumers can use to exert control over the producer (Foxall, 1988, pp. 242–3). This is highlighted by Langlois and Cosgel (1998, p. 110) in terms of consumption requiring a set of routines which can form distinctive capabilities for the firm. These capabilities not only include better communication of a firm's needs to suppliers' capabilities (Langlois and Cosgel, 1998, p. 112), but perhaps more fundamentally identifying and articulating what these existing and new needs are to itself (Robertson and

Wu, 2001, p. 190). For firms producing such goods and services, developing better customer knowledge competencies are essential (Li and Calantone, 1998, pp. 26–7).

This links in with the lack of a more holistic view of consumption within the firm and how it relates to innovation. However this now may be changing. It has been recognized that purchase decisionmakers and product users are often not the same group within a firm (Pae et al., 2002, p. 720) and that lack of adequate linkages between purchasers and users within an organization can lead to poor purchasing decisions with regard to new goods and services. A fractured and departmentalized process for the buying, use and more general consumption of goods and services can lead to lost opportunities in relation to harnessing a firm's potential capabilities in terms of consumption. Firms need a more integrated consumption knowledge framework which they can harness in distinctive ways to form a core capability of the firm.

5. CONCLUSIONS

This chapter has sought to explore consumption and services in the innovation process. It has also highlighted the important combinatorial role of services in goods consumption and more specifically how this may influence and shape the innovation process. This study, using process of encapsulation as a lens to view these various relationships has therefore attempted a more distinctive approach towards the innovation process within services and one which seeks to emphasize the role of consumption in understanding this process. In particular, it has sought to unpack the issue of consumption and how this is associated with the interrelationship between goods and service innovations.

In so doing it has highlighted that consumption as a process is not a narrow activity solely focused around simply buying or using, but instead should be considered as a much wider process and activity. The problem for both academics conceptualizing the process and for firms themselves is how wide should they conceive the process to be. For firms themselves, part of the reason why they have not been able to exploit their position as consumers is that they have considered different elements of the consumption process (buying, adoption internally, use, modification and socialization, etc.) as disparate and unconnected elements. As such, they have not been able to leverage more out of their position as consumers or fully develop their consumption capabilities in relation to the innovation process.

NOTES

1. Thanks go to executives and managers from the following companies for spending time discussing and developing these issues: Abbey National, Alstom, AstraZeneca, Ford, Pacific Power and Rolls-Royce. Thanks also go to comments made by various members of CRIC's consumption and services teams. The views expressed are the author's alone.
2. Indeed this has been noted more generally elsewhere by Randles (2001, p. 19), namely that '. . . there is an enduring tendency to conflate "consumption" with "final consumption" and attach this firmly (and therefore partially) at the sight of domestic and household consumption.'
3. This is echoed in an earlier work by George Katona entitled *The Powerful Consumer* where he talks about habitual problem solving frameworks which steer decisions in a certain direction (Katona, 1960, pp. 58–9).
4. Although as Joan Robinson (1962, p. 48) noted in her analysis of neoclassical theory of utility: 'Utility is a metaphysical concept of impregnable circularity; utility is the quality in commodities that makes individuals want to buy them, and the fact that individuals want to buy commodities shows that they have utility.'
5. Thus in the case of GE Aircraft Engines, engines can be rented from a period as short as 24 hours up to a year or more in length. Similarly GE offers Hourly Billed Services in its Medical Systems division.
6. Service encapsulation is not necessarily a new phenomenon. The example of Xerox and the sale of its reprographic equipment is more than 30 years old. Encapsulation in this context was less an intentional act, arising from careful strategy, but rather arose out of necessity. Xerox quickly realized that its reprographic machines were too expensive and indeed too unreliable (requiring high maintenance levels), for target customers to buy them outright. Instead, Xerox signed up customers with leasing contracts which gave customers varying use, maintenance and service levels. Encapsulation could also be conceived as taking place earlier in the computer industry when computer hardware manufacturers, such as IBM and Honeywell, sold hardware bundled with free software. This changed, however, in 1969, when a court ruling required IBM, under anti-trust pressure from the US Department of Justice, to unbundle its hardware and software services. As a result, IBM could no longer sell computer hardware with free software 'bundled' and a single price for the package.

REFERENCES

Bettencourt, L. A., A. L. Ostrom, S. W. Brown and R. J. Roundtree (2002), 'Client co-production in knowledge-intensive business services', *California Management Review*, **44** (4), 100–28.

Bianchi, M. (1998), 'Taste for novelty and novel tastes: the role of human agency in consumption', in M. Bianchi (ed.), *The Active Consumer*, London: Routledge, pp. 64–86.

Bressand, Alain (ed.) (1986), Europe in the New International Division of Labor in the Field of Services, final Report to FAST II, Directorate General for Science, Research and Development (DGXII), Brussels: Commission of the European Communities.

Bressand, A. and K. Nicolaidis (eds) (1989), *Strategic Trends in Services*, New York: Ballinger.

Britton, S. (1990), 'The role of services in production', *Progress in Human Geography*, **14** (4), 529–46.

Burt, D. N. and W. R. Soukup (1985), 'Purchasing's role in new product development', *Harvard Business Review*, **63** (5), 90–7.

Cohen, W. and R. Levinthal (1990), 'Absorptive capacity: a new perspective on learning and innovation', *Administrative Science Quarterly*, **35** (1), 128–52.

Cooper, L. G. (2000), 'Strategic marketing planning for radically new products', *Journal of Marketing*, **64**, 1–16.

Daniels, P. W. and J. R. Bryson (2002), 'Manufacturing services and servicing manufacturing: knowledge-based cities and changing forms of production', *Urban Studies*, **39** (5/6), 977–91.

De Brentani, U. (1995), 'New industrial service development: scenarios for success and failure', *Journal of Business Research*, **32**, 93–103.

Earl, P. E. (1986), *Lifestyle Economics: Consumer Behavior in a Turbulent World*, Brighton: Wheatsheaf.

Foxall, G. R. (1987), 'Strategic implications of user-initiated innovation' in R. Rothwell and J. Bessant (eds), *Innovation: Adaptation and Growth*, Amsterdam: Elsevier, pp. 25–36.

Foxall, G. R. (1988), 'The theory and practice of user-initiated innovation', *Journal of Marketing Management*, **4** (3), 230–48.

Gadrey, J. F., O. Gallouj and O. Weinstein (1995), 'New modes of innovation: how services benefit industry', *International Journal of Service Industry Management*, **6** (3), 4–16.

Gershuny, J. I. (1978), *After Industrial Society: The Emerging Self-Service Economy*, London: Macmillan.

Gershuny, J. I. and I. D. Miles (1983), *The New Service Economy*, London: Pinter.

Greenfield, H. I. (1966), *Manpower and the Growth of Producer Services*, New York: Columbia University Press.

Gualerzi, D. (1998), 'Economic change, choice and innovation in consumption', in M. Bianchi (ed.), *The Active Consumer*, London: Routledge, pp. 46–63.

Hill, P. (1977), 'On goods and services', *Review of Income and Wealth*, **23** (4), 315–38.

Holt, K. (1987), 'The role of the user in product innovation', in R. Rothwell and J. Bessant (eds), *Innovation: Adaptation and Growth*, Amsterdam: Elsevier, pp. 1–12.

Howells, J. (2000), 'Innovation and services: new conceptual frameworks', CRIC discussion paper 38, University of Manchester.

Howells, J. (2001), 'The nature of innovation in services', in *Innovation and Productivity in Services*, Paris: OECD, pp. 55–79.

Howells, J. (2004), 'Innovation, consumption and services: encapsulation and the combinatorial role of services', *The Services Industries Journal*, **24**, 19–36.

Katona, G. (1960), *The Powerful Consumer: Psychological Studies of the American Economy*, New York: McGraw-Hill.

Lancaster, K. J. (1966), 'A new approach to consumer theory', *Journal of Political Economy*, **74** (2), 132–57.

Lancaster, K. J. (1971), *Consumer Demand: A New Approach*, New York: Columbia University Press.

Lancaster, K. J. (1991), *Modern Consumer Theory*, Aldershot, UK and Brookfield, USA: Edward Elgar.

Langlois, R. N. (2001), 'Knowledge, consumption, and endogenous growth', *Journal of Evolutionary Economics*, **11** (1), 77–93.

Langlois, R. N. and M. M. Cosgel (1998), 'The organization of consumption', in M. Bianchi (ed.), *The Active Consumer*, London: Routledge, pp. 107–21.

Li, T. and R. J. Calantone (1998), 'The impact of market knowledge competence on new product advantage: conceptualisation and empirical examination', *Journal of Marketing*, **62** (4), 13–29.

Loasby, B. J. (1998), 'Cognition and innovation', in M. Bianchi (ed.), *The Active Consumer*, London: Routledge, pp. 89–106.

Lundvall, B. (1992), 'User–producer relationships, national systems of innovation and internationalization', in B. Lundvall (ed.), *National Systems of Innovation: Towards a Theory of Innovation and Interactive Learning*, London: Pinter, pp. 349–69.

Marshall, A. (1899), *Elements of Economics of Industry*, 3rd edn, London: Macmillan. (1st edn, 1892).

Mathé, H. and R. D. Shapiro (1993), *Integrating Service Strategy into the Manufacturing Company*, London: Chapman & Hall.

Metcalfe, J. S. (2001), 'Consumption, preferences, and the evolutionary agenda', *Journal of Evolutionary Economics*, **11** (1), 37–58.

O'Farrell, P. N. (1995), 'Manufacturing demand for business services', *Cambridge Journal of Economics*, **19** (4), 523–43.

O'Farrell, P. N. and L. A. R. Moffat (1991), 'An interaction model of business service production and consumption', *British Journal of Management*, **2** (4), 205–21.

Pae, J. H., N. Kim, J. K. Han and L. Yip (2002), 'Managing intraorganizational diffusion of innovations: impact of buying center dynamics and environments', *Industrial Marketing Management*, **31** (8), 719–26.

Parkinson, S. T. (1982), 'The role of the user in successful new product development', *R&D Management*, **12**, 123–31.

Penrose, E. (1995), *The Theory of the Growth of the Firm*, 3rd edn, Oxford University Press, Oxford (1st edn, 1959).

Petit, P. (1986), *Slow Growth and the Service Economy*, London: Pinter.

Quinn, J. J. and K. Dickson (1995), 'The co-location of production and distribution: emergent trends in consumer services', *Technology Analysis and Strategic Management*, **7** (3), 343–54.

Randles, S. (2001), 'On economic sociology, competition and markets: sketching a neo-polanyian IEP framework' *mimeo*, CRIC, University of Manchester.

Richardson, G. B. (1972), 'The organisation of industry', *Economic Journal*, **82** (327), 883–96.

Robertson, P. L. and T. F. Yu (2001), 'Firm strategy, innovation and consumer demand', *Managerial and Decision Economics*, **22** (4/5), 183–99.

Robinson, J. (1962), *Economic Philosophy*, London: Penguin.

Ruffles, P. C. (2000), 'Improving the new product introduction process in manufacturing companies', *International Journal of Manufacturing Technology and Management*, **1**, 1–19.

Sampson, G. P. and R. H. Snape (1985), 'Identifying the issues in trade in services', *The World Economy*, **8**, 171–81.

Saviotti, P. P., and J. S. Metcalfe (1984), 'A theoretical approach to the construction of technological output indicators', *Research Policy*, **13** (3), 141–51.

Scitovsky, T. (1992), *The Joyless Economy: The Psychology of Human Satisfaction* revised edn, Oxford: Oxford University Press.

Shaw, B. (1985), 'The role of the interaction between the user and the manufacturer in medical equipment innovation', *R&D Management*, **15**, 283–92.

Shaw, B. (1987), 'Strategies for user-producer interaction', in R. Rothwell and J. Bessant (eds), *Innovation: Adaptation and Growth*, Amsterdam: Elsevier, pp. 255–66.

Shostack, G. L. (1977), 'Breaking free from product marketing', *Journal of Marketing*, **41** (2), 73–80.

Silvestro, R., L. Fitzgerald, R. Johnston and C. Voss (1992), 'Towards a classification of service processes', *International Journal of Service Industry Management*, **3** (3), 62–75.

Stigler, G. J. and G. S. Becker (1977), 'De gustibus non est disputandum', *American Economic Review*, **67** (2), 76–90.

Swann, G. M. Peter (1999), 'Marshall's consumer as an innovator' in S. C. Dow and P.E. Earl (eds), *Economic Organization and Economic Knowledge: Essays in Honour of Brian J. Loasby*, Cheltenham, UK and Brookfield, VT: Edward Elgar.

Thomas, R. (1994), 'Purchasing and technological change: exploring the links between company technology strategy and supplier relationships', *European Journal of Purchasing and Supply Management*, **1** (3), 161–8.

Toffel, M. W. (2003), 'The growing strategic importance of end-of-life product management', *California Management Review*, **45** (3), 102–29.

von Hippel, E. (1976), 'The dominant role of users in the scientific instrument innovation process', *Research Policy*, **5** (3), 212–39.

von Hippel, E. (1978), 'Successful industrial products from customer ideas', *Journal of Marketing*, **42** (1), 39–49.

von Hippel, E. (1988), *The Sources of Innovation*, Oxford: Oxford University Press.

von Hippel, E. (1994), ' "Sticky information" and the locus of problem solving: implications for innovation', *Management Science*, **40** (4), 429–39.

Witt, U. (2001), 'Learning to consume – a theory of wants and the growth of demand', *Journal of Evolutionary Economics*, **11** (1), 23–36.

Woo, H. K. H. (1992), *Cognition, Value, and Price: A General Theory of Value*, Ann Arbor, MI: University of Michigan Press.

Wood, P. A. (1991), 'Flexible accumulation and the rise of business services', *Transactions of the Institute of British Geographers*, **16**, 160–73.

Wood, P. A. (1996), 'Business services, the management of change and regional development in the UK: a corporate client perspective', *Transactions of the Institute of British Geographers*, **21**, 649–65.

Wynstra, F., B. Axelsson and A. van Weele (2000), 'Driving and enabling factors for purchasing involvement in product development', *European Journal of Purchasing and Supply Management*, **6** (2), 129–41.

Wynstra, F., M. Weggeman and A. van Weele (2003), 'Exploring purchasing integration in product development', *Industrial Marketing Management*, **32** (1), 69–83.

PART TWO

Innovation and firm strategy

4. Paths to deepwater in the international upstream petroleum industry

Virginia Acha and John Finch[1]

1. INTRODUCTION

This chapter presents an analysis of recent developments and adaptations of capabilities among a subset of companies active in the world's upstream oil and gas industry. These changes are due to this group of companies undertaking exploration and also production of hydrocarbon resources offshore, in deepwater. The change is profound, and recognized among industry participants in which 'deepwater' has become a term of industry-wide significance describing a category of activity. Our study focuses partly on path dependencies in this development. However, interactions among oil companies (or operating companies) and services and supply companies are crucial, so paths are partly at the industry level, and partly in terms of identifiable approaches of individual companies. Despite deepwater being an object of investigation and research among some companies since the 1970s, a significant accumulation of activity has been ongoing since the mid 1990s.

Informal industry standards recognize deepwater exploration and production as occurring in water depths of over 400 metres, and ultra-deepwater exploration and production activities in water depths of over 1000 metres. The very currency of 'ultra-deepwater' emphasizes that technological developments are ongoing, and that definitions lag behind. Further, despite oil companies acquiring considerable exploration rights from governments in licensing rounds going back to the early 1990s (termed 'acreage'), and despite hundreds of exploration wells being drilled, including much publicized 'record depths,' the numbers of fully developed projects that are now in production is still relatively small. This pattern draws attention to the sequential decisionmaking procedures of senior managers, to the long timeframes involved in both exploration activity, and to the attendant decision analysis and decisionmaking procedures within oil companies.

Our analysis is developed along three critical, and interlinked, dimensions: (1) the development of exploration capabilities among geoscientists in the upstream industry; (2) the development of engineering capabilities in both exploration and production; and (3) the organization of these capabilities, seen both in industry terms, and in terms of decision analysis and decisionmaking procedures of oil companies. Drawing from mainly theoretical work on capabilities, and the organization of industrial activities, we contend that companies in the industry have had to adapt technologies, and also organizational patterns, that are well suited to other forms of exploration and production activities (by default, 'shallow water'). These had become increasingly routine and standardized during the 1970s and 1980s, so much so that major oil companies had been selling such production activities (or assets) to smaller oil companies, many of which used such acquisitions to enter particular areas, such as the North Sea. To preempt our detailed analysis, the major oil companies involved in deepwater exploration and production are exhibiting characteristics of intransigence in maintaining and developing their capabilities in-house or close at hand. Crucially, and despite the significance of deepwater as a global category of activity, and attendant industrywide initiatives, the oil companies are doing this differently.

Our theoretical concerns address the application of the capabilities approach, usually traced to Penrose (1959) and Richardson (1972). Loasby (1998, 1999) draws these together in setting out capabilities as being complexes of knowledge, with some elements articulated in code, and others remaining tacit, that are held within companies, or close at hand from a particular company's perspective. Further, we recognize in the capabilities perspective that companies, under the direction of senior managers, normally undertake a range of activities. Penrose's crucial contributions are, arguably (1) in establishing managerial discretion in directing or configuring what she terms 'resources' towards one set of activities, bringing personal knowledge of entrepreneurial opportunities to bear; and also (2) in setting out how companies possess the innate capability of embedding routines and delegating or distributing what were once novel activities and their attendant capabilities, freeing up managerial attention and resources to plan and develop new activities. Richardson's (1972) key contribution is in addressing the wider organization of activities within what we may term industrial networks. Hence, activities require sets of complementary capabilities, yet these may be similar or dissimilar in terms of the skills and knowledge required for their performance. Richardson (1953) makes an earlier contribution by distinguishing, essentially, between personal knowledge and declarative knowledge. He argues that business decisionmaking can be based on analysis that is inalienable from an agent, or one that is objective in the sense that all individuals with similar data and methods can arrive at the same answer.

The ideas have been developed further, and not exclusively within the capabilities tradition, by Langlois (1992, 2002). Langlois (1992) argues that the boundaries of the firm are shaped by the relative learning capability of that firm, compared with potential competitors, collaborators and suppliers in their industry grouping. Langlois (2002) puts forward a 'modular theory of the firm', arguing that firms tend towards an equilibrium characterized by monocapability in which decision, residual and property rights coincide within a top management team (see also Richardson, 2002).

This chapter is structured as follows. Section 2 sets the stage for our discussion with a brief overview of the evolution of deepwater activity. Section 3, the theoretical framework in which we will couch our argument, follows. In Section 4 we apply our framework to case studies of the deepwater activities of three major international oil companies, and we identify regularities (and some irregularities) across these companies as they relate to the framework. We conclude by drawing together the main ways in which companies shift capabilities and governance structures in a turbulent and uncertain environment.

2. DEEPWATER: HUNTING THE LAST ELEPHANTS

As we begin the 21st century, oil companies have moved beyond what has been conventional onshore and offshore exploration and production and into deeper waters in search of the last remaining 'elephant fields'.[2] Industry statistics testify to the dwindling number of major oil and gas finds to be made. In the 1960s, average annual discoveries worldwide comprized 70 billion barrels of oil equivalent (boe); by the 1990s the annual average was only 20 billion boe. Likewise, the size of fields (in terms of volume) found is shrinking, from over 200 million boe per discovery in the 1960s to less than 50 million boe in the 1990s (Hayward, 1999, p. 42). Moreover, we know that these sliding figures are not due to the lack of efficiency or effort on the part of oil companies. In fact, the strike rate (the ratio of producing to dry wells drilled) has doubled from one in ten in the 1960s to one in five in the 1990s and yearly investment in new fields amounts to several billion USD. The argument is that almost all of the major oil and gas finds in the conventional onshore and offshore regions have been found and exploited. Many commentators have warned that the industry was fast approaching 'Hubbert's peak' at the end of the millennium.[3] With the improvements in seismic technologies to probe previously inaccessible sites, perhaps the last remaining 'elephants' have been identified at the bottom of the oceans.

The first offshore exploration and production took place early in the oil and gas industry's history. In 1887, a Californian was encouraged by good production records on nearshore wells to build a wharf and erect a drilling rig. Success from this venture led to several other wharves stretching into the Pacific. However, the expansion of offshore production really took off in the 1950s, following the first out-of-sight-of-land fixed platform well in 1947, which was drilled by Brown & Root (an oil services company) with and on behalf of Kerr-McGee (the oil operator). Several other companies quickly followed this pathbreaking effort: 'By 1949, 11 fields were found in the Gulf of Mexico with 44 exploratory wells' (NOIA, undated). By the end of the 1960s, there were over 800 platforms in use in the Gulf of Mexico alone. This expansion into offshore was checked somewhat by the collapse in oil prices after 1984, when many planned exploration projects were shelved. However, the advance of technologies for offshore (in particular seismic technologies and remotely operated vehicles) made it possible for companies to move further out from shore and into greater water depths.

Deepwater oil and gas plays are in five main regions: offshore Brazil (the Campos basin), the Gulf of Mexico, the North Sea, offshore Western Africa and South-east Asia. The North Sea and South-east Asian provinces are largely gas-bearing, whereas the other three provinces are predominantly oil-bearing provinces (Shirley, 2002).[4] The first deepwater efforts were made in the 1970s (at water depths greater than 500 metres), but the first significant successes occurred in the 1980s with the development of fields in deepwater Campos Basin (by Petrobras) and in the Gulf of Mexico (by Royal Dutch Shell). Hence, 'Since then, success rates and volumes discovered have improved significantly with 20 per cent of the world's oil discoveries and 6 per cent of the world's gas finds made in deepwater during the last decade' (Hayward, 1999, p. 43). Improvements in seismic imaging and advances in well engineering have made deepwater developments less expensive and less risky over time. The economics of deepwater have also been demonstrated to be favourable because, despite the relatively high costs of development in deepwater, production volumes are relatively large.

> The term 'productive' has, of course, a special meaning in deepwater. Economic development requires wells capable of producing over 20 million barrels at rates above 10 000 barrels a day. Such 'high ultimate-high rate' wells are the first imperative for deepwater profitability. (Watts, 1998, p. 2)

Not all oil companies are active in deepwater exploration and production. The first efforts were made by Petrobras and Royal Dutch Shell (henceforth, Shell), but these companies were soon joined by Exxon, BP, Elf (notably in West Africa) and Mobil, and later by Chevron, Texaco and Arco. Petrobras began the development of the deepwater Campos Basin in Brazil in 1977, reaching world records for water depths in the 1980s and 1990s. In the Gulf of Mexico, the Shell Cognac field was a milestone in that it broke the 1000 feet water depth record in 1978, followed by the Exxon Lena field in 1984, production at which confirmed the viability of the deepwater reservoirs (Godec et al., 2002). In the North Sea, both Shell and BP, faced with maturing existing fields, extended field development plans to the deepwater West of Shetland province. BP began to explore in waters beyond 500 metres in the mid 1970s, but it took nearly 20 years before the first viable discovery was made in 1992, the Foinaven field, followed soon after with the discovery of Schiehallion and Loyal. Likewise, Shell began its exploration of deepwater opportunities in the form of turbidites in the North Sea and the Gulf of Mexico in the 1960s (Watts, 1998). Of course, it must be emphasized that throughout the North Sea development, Shell and BP (and to a lesser extent Elf Aquitaine) were gaining considerable expertise in exploring and producing in harsh environments at deeper and deeper depths. 'Some striking Shell publicity posters in the 1970s likened the North Sea development to space exploration' (ibid., p. 1).

Interest in deepwater exploration and production developed significantly in the mid 1990s following the successful development of the Auger field by Shell in 1994. Auger was unique because it was the first Tension Leg Platform and the field was developed at a world water depth record of 2860 feet. The Auger development was a demonstration of the viability of economical deepwater exploration and development. From that point forward, the numbers of oil operators active in deepwater plays grew quickly and continues to do so.

In the five years from 1998 to 2002 deepwater reserves totalling some 10.6 billion barrels of oil equivalent (boe) were brought onstream by 14 oil company operators. In contrast, over the 2003 to 2007 period, 37 operators have deepwater prospects targeting almost 32.4 billion boe (Infield, 2002).

Alongside operators, the service companies and drilling contractors were gaining capabilities and focusing their services on the deepwater market. Transocean, a leading drilling contractor, set many deepwater drilling records in the 1970s, the 1980s and the 1990s (Rose, 1999). All of the 'major' suppliers, such as Schlumberger, Halliburton and Baker Hughes, have amassed broad portfolios of services and products dedicated to deepwater exploration and production. Specialist designs for deepwater drilling platforms, namely Tension Leg Platforms, floating production systems and Spars, have been developed by the supplier companies to serve the deepwater expansion. Specialist deepwater suppliers have also emerged, particularly with respect to new subsea system designs and components.

What makes the deepwater movement different from the original move to offshore production is the slower pace of diffusion and imitation of activities and technologies. From the Kerr-McGee well to the next ten wells in the shallow depths of the Gulf of Mexico, there was but a few years. In fact, the first Texas offshore well was drilled in 1942, only five years prior to the Kerr-McGee development. The move to deepwater has taken decades of investment, research and exploration by the leading oil operators and Petrobras, and the take-up of the deepwater model of exploration and production has really only taken place since the mid 1990s, after the 'proof of concept' by Shell through the Auger development (which followed several other Shell deepwater projects in the Gulf of Mexico). The reasons for this slower cycle is most likely due to the high capital risk involved as well as the considerable technical risk of operating at such water depths. These technical risks have required not only adaptation of existing technologies to deepwater environments, but also entirely new technologies and techniques such as new platforms, subsea production systems, 'smart' well completions (that allow remote monitoring), advanced drill ships and the like. As the Group Managing Director for Shell argued:

> I have stressed the extent to which deepwater advances are the product of long-term technological evolution. Floating production systems have been developed over two decades. The necessary reliability for deepwater subsea systems was learned in shallower waters. But there's more to it than that . . . Over the past 20 years Shell companies have invested over $500 million in deepwater research – experimenting, for instance, with platform, pipeline and riser designs in our own test tanks. (Watts, 1998, p. 5)

3. TRANSLATABLE CAPABILITIES, RIGIDITIES, AND NEAR MATCHES

In this section, we provide a theoretical framework in which the observed tendencies of capability development in the industry may be interpreted and explained. Given the dynamic, innovative and organizational (internal and external to particular companies) implications of deepwater exploration and production, the capabilities approach seems to be a reasonable starting point for developing such a theoretical framework (Loasby, 1998, 1999). In particular, we focus on two overlapping strands of this approach: a Penrosian-cum-Chandlerian version that emphasizes the tenacious and intransigent qualities of companies as ongoing organizations, or as stores of potential business development; and one that draws from Richardson

(1972) and Langlois (1992, 2002) that emphasizes the coordinating imperatives of requiring both complementary similar and dissimilar capabilities in order to undertake complex business activities.[5] Our overriding task though is to provide a framework or explanation that is in accordance with our observations of deepwater activity in which oil companies are devoting resources, seemingly with the effect of keeping equilibrium tendencies – in terms of firm-level and industry-level analysis – at bay.

The embedding of routines within companies is key to both of the capabilities strands. For Penrose, the embedding of routines creates additional resources, which in the first instance are located within a particular company. Beyond this, a company has to be growth-oriented; otherwise it can just get on with becoming a routine-led company, with the expectation of normal profits.[6] Growth orientation implies that routinization frees up resources for internal and external search, developing new business, and integrating this new business into established activities. Path dependence is of critical importance because search and integration are shaped by managerial experience, and what Penrose describes as resources (which are in her sense practically indistinguishable from Richardson's capabilities) also develop in contexts. Resources close at hand to managers, such as their own personal knowledge, and those available within the organization, through routinization or through redirecting activities, provide a constraint on otherwise imaginative interpretations of business experience, hence, Leonard-Barton's (1992, 1998) description of 'core rigidities'. This may however better be described as a dialectic of routines and of imagination, or of codified, articulated or otherwise shared routines and practices, accompanied by visions and ways of knowing as yet inalienable from individuals.

Further, as Chandler (1990) argues, organizational capabilities (similar to routines) require maintenance. The point about managerial experience, which implies that Penrose had in mind personal, subjective and inalienable knowledge, and resource development 'close at hand', is particularly important among that subset of oil companies prominent in deepwater exploration and production. These multinational and vertically integrated companies often act as internal selection environments for seeking potential senior managers. Hence, an internal career path is established among these companies where potential senior managers gain experience within the company, but in different projects, geographical locations and roles.

Following this line of argument, managers have, in Chandlerian terms, a crucial role in organizing, and in maintaining means of organizing, though establishing means by which distributed knowledge and expertise can be coordinated. In Penrosian terms, managers also have a key role in imagining and deciding how to expand and diversify through undertaking new business. Search in these terms includes drawing on personal knowledge, developed in a career, and filtering and selecting between alternative modes of, or plans for, expansion and diversification. Indeed, a company's organizing capabilities provide a structure by which entrepreneurial, and so subjective, plans may be shared, assessed and adapted. In these terms, the problem that oil companies have in expanding their activities into deepwater and ultra-deepwater exploration and production involves matching established routines in decision analysis and other operations (whether formalized into codified systems or not) with the characteristics of new business.

From Richardson's (1972) perspective, the problem faced by companies in the upstream petroleum industry is coordinating dissimilar capabilities that are nevertheless highly complementary in the context of particular activities; in this case, offshore exploration and production. Richardson's capabilities are a little different from Chandler's, in that Richardson seems to be focusing on discrete tasks, such as geoscientists analysing seismic data, or engineers designing exploration and production wells. Chandler's capabilities pertain to the organization itself and are involved in the coordinating process, with respect to different corporate functions. Richardson's advice to company managers is that similar capabilities can be coordinated effectively within one organization, but that different companies developing close working relations should coordinate dissimilar capabilities. The context of similarity is provided by an activity. As Langlois and Robertson (1995) have argued, 'bottlenecks' and 'anti-bottlenecks' can be more effectively handled within organizations combining similar capabilities, so, for example, specialist geoscience consultancies can develop business relations with many oil companies.

Richardson's (1972) distinction between complementary activities that are either similar or dissimilar is of critical importance to understanding the capabilities approach as it focuses on corporate boundaries. The distinction seems clear cut. In the case of the upstream petroleum industry, the complementary capabilities may include seismic analysis, geoscience, engineering and economic evaluation. These are all different activities and can be identified as such through participants having distinct professional training, manifested in professional organizations. However, the context of complementarity adds an important qualification. Complementarity is in the context of the overall activity, and could otherwise be called context-dependent, such as upstream petroleum exploration. The fact of shared context invites a more detailed examination of how we may categorize capabilities as similar or dissimilar. A problem faced by oil companies is not so much managing dissimilar complementary activities, but in creating conditions of complementarity itself. And once this is created, the similarity seems to develop. Hence, there are working groups where membership includes petroleum geoscientists, and petroleum engineers, and economists.

Our argument is then that similarity and dissimilarity are relative terms for capabilities that can be understood in the context of complementarity and in practice. And in practice, if managers are to plan and organize capabilities in the manner set out by Richardson, they need some way of recognizing and instituting similarity and dissimilarity in context. Any separation among now identifiable similar or dissimilar capabilities required in undertaking activities requires accompanying codification, perhaps through company operating procedures, and, if involving separate companies, corporate contracts (Ancori et al., 2000; Cohendet and Meyer-Krahmer, 2001; Cowan et al., 2000; Cowan, 2001). In effect, this highlights the Chandlerian notion of organizing capabilities, alongside, and complementary with, the 'direct' capabilities required in the upstream petroleum industry, such as seismic interpretation, geoscience, engineering and economic analysis. And Chandler's organizing capability, when recognized as such, is a dissimilar capability.

If the identification of capabilities, and indeed formulating these in the context of activities and complementarity, is a dynamic and emergent process, we can examine these further in the contexts of routines and of changing company boundaries. Langlois (1992, 2002) sets out an argument in which firms tend towards equilibrium characterized by, in Richardson's (1972) terms, monocapability. Langlois (1992) argues that managers experience dynamic transaction costs in acting upon the otherwise seemingly autonomous process of routinization, and these are mainly the costs of codification or other articulation in transferring activities, and attendant (now) dissimilar capabilities, to other organizations. Dynamic transaction costs may also include coping with the consequences of a rival firm imitating such capabilities. Langlois (2002) develops another line of argument, in terms of firms tending over time to become modular organizations, which contain well aligned residual rights in property, decisionmaking, and revenues. Hence, modules within firms, as discussed in the context of production technology and product design by Sanchez and Mahoney (1996), are not truly modular, as rights are not well aligned inside these organizational structures.

Langlois's arguments, though, can be interpreted as identifying and isolating tendencies in the dynamic organization of business activities. While in Langlois (1992) he outlines an inimitable core of capabilities, it could be argued that designation of inimitability is arbitrary. Again referring to recent work on codification of knowledge, and of communities of knowing, inimitability depends on the willingness of participants in an industry to devote resources to codification. Hence, knowledge is neither tacit nor codified, but emerges in particular forms in contexts as a way of knowing (Brown and Duguid, 2001). Following Penrose (1959), we argue that companies have additional tendencies of resilience and tenacity in the face of the tendencies outlined by Langlois. These may not be manifest evenly across an organization, perhaps appearing among particular operating groups, or divisions, or among senior managers. Such new activities may well be sources of corporate renewal, and, returning to the discussion of Richardson and similar and dissimilar capabilities, start to blur boundaries in context, awaiting later codification if not modularization. Indeed, this has something in common with Richardson's (1953) discussion of personal and declarative knowledge, of individuals possessing unique ways of envisaging, planning and assessing possible courses of action, as opposed to those drawing upon more readily shared approaches.

Further, and given the introduction of organizational capabilities in the Chandlerian sense, it is not obvious that capabilities can be organized, disentangled, and recognized for the purposes of corporate development, at the same rate. In the case of deepwater exploration and production, geoscience and seismic analysis have been at the forefront of establishing this sector within the industry. The analysis of three dimensional seismic representations has given geoscientists much greater confidence in identifying significant hydrocarbon prospects in deepwater and ultra-deepwater. But that subset of companies in the industry who have undertaken deepwater exploration, and also in most cases at least some production, have to cope with different sets of organizational contingencies, in coping with the now routine activities of managing mature assets, and also freeing up established and perhaps otherwise underused exploration capabilities to develop proposals of deepwater exploration and production.

Deepwater and ultra-deepwater exploration and production have provided an impetus to further technological developments in the upstream petroleum industry. Some activities have benefited from adapting existing technologies:

> Should new purpose-built rigs be constructed at high cost and high saturation risk? Yes, if necessary, but finding the means to modify what is available through economically innovative methods achieves greater economic efficiencies for large deepwater shareholders such as platform owners and operators. (Smith et al., 2001, p. 37)

Alternatively, publicity has been generated around the concept of fully automated drilling rigs located on seabeds:

> A three legged unit operated entirely remotely, complete with drilling derrick and pipe store, mud pumps, etc . . . could operate in water depths of up to 3000m in all deepwater regions, such as West Africa, Brazil, the Gulf of Mexico, and even under ice in Arctic sectors. . . . Pilot work on the concept began whilst a

highly-automated rig was designed for the Troll field in Norway. Following this, in 1997, Shell Research and Technology's Gamechanger panel . . . approved funding for a further study of a subsea rig concept. The concept is still being taken seriously, with . . . the JIP [joint industry project] set up between Shell and Saipem. (Thomas and Hayes, 1999, p. 35)

What is clear is that deepwater conditions have required a rethink in exploration and production activities, be these in terms of recombining existing techniques and equipment within different overall approaches, or in designing radically different equipment within the context of different overall approaches. Some capabilities have proved translatable from shallow water to deepwater contexts, whereas others are at best 'near matches'. For example, for the most part deepwater down-hole well completions are conventional (Moritis 2000), but some materials and techniques (for example, flow assurance) have to be altered to meet the high pressure, high temperature environments of deepwater. 'Smart' well completions are completely new to the industry. As subsea production becomes more commonplace, 'tree hugging' oil operators (companies that prefer topside production controls) will have a steeper learning curve to face than operators that have long since adopted 'wet tree' methodologies.[7]

At the risk of oversimplification, deepwater has required a recombination of industry expertise along the dimensions of: geoscience and seismic interpretation, engineering and equipment design, and in decision analysis and decisionmaking, especially with respect to risk, uncertainty, and economic and financial consequences of committing a portion of an oil company's resources to deepwater activity.

4. THREE DEEPWATER CASE STUDIES

This section is based on case studies of three major oil companies, Arco, BP and Statoil, which have undertaken deepwater exploration and production activities.[8] These companies vary in size, the extent to which they are multinational companies in coverage of deepwater locations, whether they are national companies (with significant state share ownership), and in terms of company size and performance. While companies may have similar motives to undertake deepwater exploration and production, in requiring new activities in which their established capabilities could be employed with some significant adaptation, different internal and external organizational means were pursued. We explore the patterns of participation in deepwater of the three companies and, informed by the capabilities approach, consider what rationales lie behind these patterns.

Before addressing the firm-specific influences, we must also recognize that external influences have also shaped the pattern of participation in deepwater of these three companies. Governments, as licensing agencies, have shaped deepwater exploration in their provinces differently. So companies with established presences in, for example, the Gulf of Mexico, the UK's Atlantic margin, the Norwegian Sea's Atlantic margin, the West Coast of Africa and Brazil, have faced different types of incentives in undertaking their deepwater activities. Where national governments have adopted managerial strategies in developing offshore hydrocarbon resources, this has affected the funding and organization of research and development. In other environments, joint industry projects have been established among companies, most prominently in the Gulf of Mexico with the Deep Star project. These variations in adapting existing capabilities for deepwater activities are critical because, in geological terms, a range of similar exploration and production problems are raised, irrespective of the location of that activity.

The very presence of these industry and multifirm organizational patterns highlights the problems faced in coordinating activities in both mature types of field developments, and in the deepwater and ultra-deepwater types. An organizational pattern consistent with Langlois' notions of dynamic transaction costs and also the modular theory of the firm fit much better with oil companies' managing of mature assets, than with bringing together capabilities – some as nascent forms – for deepwater activities.

As stated earlier, BP has been an early investor in deepwater capabilities from the 1970s through its participation in the North Sea and, in the 1990s, the Gulf of Mexico. At present, BP is currently holding more deepwater acreage than Shell, although Shell has the highest level of production. Atlantic Richfield Corporation (ARCO) was a latecomer amongst the oil majors to deepwater exploration, but rapidly gained prominence in the Gulf of Mexico through its deepwater arm, VASTAR: 'Vastar Resources Inc. discovered oil with the first well it drilled in the Gulf of Mexico deepwater' (Rhodes, 1998). Again, ARCO required a separate organizational form (i.e. VASTAR) in order to undertake this activity alongside its established exploration and production. Moreover, ARCO reached the position of oil major through its successful development of oilfields under Arctic conditions in Alaska; technical and financial risk had been part and parcel of that development.[9] Statoil, like BP, has accumulated significant experience in operating in the harsh weather conditions of the North Sea. Unlike BP, Statoil has not strayed much from the North Sea to operate deepwater developments in any of the other provinces, and as yet they are not active at the frontier water depths of deepwater exploration and production.

To compare these distinct patterns of participation, we return to the capabilities approach. From this, and from evolutionary economics, we would expect firms to find it difficult to change their strategy or structure quickly and easily. Path-dependent processes impose some rigidities and biases with respect to the opportunity set of choices. Because of this, we expect that translatable capabilities (in the geosciences and engineering) would facilitate a move into a new strategic direction. Finally, there must be a motive for the change in direction. Table 4.1 provides a summary of these rationales with respect to the movement of the three companies into deepwater exploration and production.

This analytical comparison of the three companies does not reflect the relative performance of the companies in deepwater activity. BP is a world leader in deepwater (together with Petrobras and Shell). ARCO had some successful deepwater discoveries through VASTAR, but had not produced from these assets before BP acquired the company. Statoil is a recent entrant into deepwater activities, particularly with respect to provinces other than the North Sea. Table 4.1 provides some clues to why these differing outcomes have emerged. Beginning with the last rationale, BP had the greatest willingness to take on the additional technical and financial risks imposed by deepwater. Because of its size and its capital structure (physical and financial), the company needs to fully utilize its capacity with an accordingly high scale of production volumes. BP cannot easily afford to operate from the basis of marginal fields, and therefore its strategy has been to focus on the large field developments. As noted in the second section of this chapter, these are primarily in the deepwater provinces.

ARCO had less incentive primarily because its Alaskan developments were the principal basis of operation for the company. Its move into deepwater was pursued to keep pace with the other major oil companies and to diversify the company's earnings structure. Statoil had less incentive as the company has long held a privileged position in the exploitation of the Norwegian North Sea province. The great majority of Norway's oil and gas resources remain in place but the technical challenges and depths in reaching it are increasing. It is also true that both Statoil and ARCO did not have the financial reserves to weather the significant financial and technological risks required of leaders in deepwater exploration and production. Of course, such financial support does not necessarily have to be within the company itself, as is demonstrated by Petrobras. Petrobras had to move into deepwater exploration and production simply because that is where the significant Brazilian hydrocarbon resources lie. The Brazilian government instigated a dynamic approach to develop the skills and capabilities internally by sending some of their best graduates abroad to study in the principal universities that specialize in oil-related teaching and research in the US and Europe. It also recruited geoscience professionals internationally, relocating personnel to Brazil to teach and pursue

*Table 4.1   Rationales for participating in deepwater exploration and
              production*

Factor	BP	Arco	Statoil
Entry to deepwater	Early mover (mid 1970s)	Late mover (mid 1990s)	Late mover (late 1990s)
Deepwater organization	Deepwater Business Development Unit, coordination of deepwater technology across business units	Separate subsidiary company: VASTAR: ARCO has a parallel deepwater technology team	Assimilated into established asset management and decision analysis procedures
Path dependency, core competencies and rigidities	Significant expertise in harsh environments, risky offshore field developments	Arctic developments (technical and financial risk); significant onshore work, well completions technology	Significant expertise in harsh environments; seismic analysis and interpretation
Translatable capabilities	Horizontal, multiwell technology; high temperature, high pressure wells; reservoir characterization	Reservoir characterization; 3D seismic processing	Reservoir characterization; multilateral wells
Exploration and production portfolio realignments and risk profiles	Requirement to find and produce only large volume fields	Diversification into other profitable exploration and production regions	Diversification and international-ization into other profitable exploration and production regions

research (Robertson, 1999). The success of Petrobras' expertise and technology in the geoscience and engineering of deepwater, particularly in the
1980s and 1990s, has been underscored by drilling milestones and technology awards.

By translatable capabilities, we address capabilities established in other
exploration and production activities that are useful in a deepwater context.

Reservoir characterization (underpinned by the geosciences) is always a core capability for an oil operator, and thus it is easily translatable to deepwater. This does not mean that novel geological structures do not occur at deepwater; this uncertainty is always at issue, and it was the creative approach of the geoscientists' team at BP that determined the unusual structure of the company's most lucrative deepwater find, Thunder Horse: 'We adopted a "back to basics" philosophy, focused on the geologic elements of that basin, while ignoring seismic attributes. This concept changed the way we view prospectivity in the GoM, and steered us towards deeper untested structures' (Yielding et al., 2002). BP holds several competitive capabilities that are translatable to a deepwater environment, including strong geology capabilities. Although the company recruited some additional specialists, the majority of deepwater expertise in BP has emerged in-house.

The path-dependent development of firm capabilities and strategy is likewise fundamental in explaining firms' strategies in entering the deepwater exploration and production sector. It is at this level that a significant contrast arises. It is perhaps easier to understand ARCO's relatively slow move (for a major) into deepwater, as the company had most of its experience in onshore field developments, albeit in the extremely risky context of the previously untested Arctic environments. ARCO was clearly a more risk-friendly company, and when it finally made the move to deepwater, it did so decisively. The contrast occurs between BP and Statoil, both of which have been earning their profits and reserves primarily in the harsh and technically difficult offshore environment of the North Sea. Whilst this corporate experience appears to have reinforced both companies' capabilities and strategy to move into deepwater, Statoil has failed to follow BP's strategy despite the company's industrywide reputation for excellent technology and engineering.

5. CONCLUDING REMARKS

We have reviewed the cases of companies involved in deepwater exploration and production in the upstream oil and gas industry. Our argument is that this subset of companies from the industry have exhibited characteristics of tenacity and resilience by continuing to maintain their exploration functions in the face of industrywide tendencies towards the routine management of mature assets in production, and reduced prospects for the discovery of further large prospects. The undertaking of deepwater exploration and production activities has created additional pressures as innovative exploration drilling and production solutions, sometimes of an incremental nature in adapting existing equipment and procedures, and sometimes requiring new equipment, are required. Further, those companies involved in deep-water exploration have faced organizational pressures in running produc-tion activities with their mature fields, alongside exploration and some production activities among their deepwater prospects.

The case of deepwater in the upstream petroleum industry provides a perspective on recent debates from those undertaking research in the context of the capabilities approach regarding the organization of eco-nomic activities. In the first instance, we identify counter-tendencies to those of routinization and modularization highlighted by Langlois. These counter-tendencies are derived from Penrose's argument that managers (in particular) within companies have the role of harnessing what are in effect resources freed up by routinization, and directing these towards new activ-ities. Further, we expect that these new activities are nonroutine, and require close working among different individuals in groups, with emergent and tacit working patterns of small grain size developing.

This nonroutine argument is consistent with Richardson's early work on personal knowledge, unique to individuals, as a means of calculation and appraisal of possible business activities. It leads onto our second theoretical point, worked out in the context of our case study analyses, in which the identification of capabilities, as indicated in Richardson (1972), becomes problematic in an explicitly dynamic framework, such as the one in which counter-tendencies are, feasibly, in operation among at least some com-panies in an industry. Whereas in principle, capabilities can be identified as being distinct, the crucial aspect in the capabilities approach is that cap-abilities are related through some context – economic activities – such that they are similar or dissimilar in this context. Similarity is not given exoge-nously, but arguably is the product of managerial and other organizational work (and autonomous processes). Furthermore, these sometimes con-scious and sometimes autonomous processes connected with routinization, and also modularization in Langlois's sense, can also include a different type of capability, such as Chandlerian organizational capability. Hence, we argue that the distinction of similar and dissimilar is in part, but signifi-cantly in part, an emergent property of the dynamic interactions between tendencies of tenacity in seeking reuse of capabilities threatened by the embedding of routines, and also of the embedding of routines itself.

Further steps are required in articulating our argument of dynamics through different tendencies. We are required to develop a means of cali-bration so as to be more precise in describing the ways that the three cap-abilities that we identify as being significant in this deepwater case (geoscience, engineering and decision analysis) are developing at different rates and in different companies. Further, we need to capture some of the processes, or perhaps nascent processes, in those larger oil companies that have not yet entered into deepwater exploration and production. We expect even nascent tendencies to have some effect as organizations develop and adapt to the maturing of their capabilities in the context of exploration and activities generally.

NOTES

1. Presented at the ASEAT/Manchester Institute of Innovation Research conference, 'Knowledge and Economic Social Change: New Challenges to Innovation Studies,' 7–9 April 2003. We are grateful to Mark Winskel, the editors of this volume and our industry participants for comments and criticisms. The usual disclaimer applies.
2. A working definition is that an elephant field comprises 100 million barrels of oil equivalent.
3. 'Hubbert's Peak' is the point of maximum production, which tends to coincide with the midpoint of depletion of the resource under consideration. Hubbert developed this analytical framework in the 1950s and many others have expanded it over the decades.
4. The US Geological Survey's (1997) definition of a play is: 'a set of known or postulated oil and (or) gas accumulations sharing similar geologic, geographic, and temporal properties, such as source rock, migration pathway, timing, trapping mechanism, and hydrocarbon type.'
5. Chandler (1962, p. 453 ff) comments on the complementarity between his own approach and that of Penrose (1959), especially her chapters on 'Inherited Resources and the Direction of Expansion,' and 'The Economics of Diversification.'
6. This 'choose growth' interpretation of Penrose (1959) led to Marris's (1966) reinterpretation of a trade-off for senior managers and shareholders. The effect of this intervention is to isolate or disembed Penrose's explanations of how growth might occur, and how it involves the dialectic of routinization and imagination, from the discretionary interventions of managers in drawing upon rational decision analysis resources to choose some optimal development path.
7. The 'tree' in this sense refers to the 'Christmas tree', which is the valve control unit, which controls the flow from a well. A 'wet tree' is a Christmas tree operating under water.
8. We draw our data from a range of sources, including: semistructured interviews with personnel of these companies, corporate annual reports, papers delivered at the conferences of the professional organizations involved in the industry, and articles form industry journals. The latter two categories of sources are included in the bibliography.
9. An 'oil major' is a joint-stock company, as opposed to a national oil company, and is involved in all stages of the industry, from exploration and production, through refining, distribution and retailing.

REFERENCES

Ancori, B., A. Bureth and P. Cohendet (2000), 'The economics of knowledge: the debate about codification and tacit knowledge', *Industrial and Corporate Change*, **9** (2), 255–87.
Brown, J. S. and P. Duguid (2001), 'Knowledge and organization: a social-practice perspective', *Organization Science*, **12** (2), 198–213.
Chandler, A. D. (1962), *Strategy and Structure, Chapters in the History of the Industrial Enterprise*, Cambridge, MA: MIT Press.

Chandler, A. D. (1990), *Scale and Scope, the Dynamics of Industrial Capitalism*, Cambridge, MA: Belknap Press.

Cohendet, P. and F. Meyer-Krahmer (2001), 'The theoretical and policy implications of knowledge codification', *Research Policy*, **30** (9), 1563–91.

Cowan, R. (2001), 'Expert systems: aspects of and limitations to the codifiability of knowledge', *Research Policy*, **30** (9), 1355–74.

Cowan, R., P. A. David and D. Foray (2000), 'The explicit economics of knowledge codification and tacitness', *Industrial and Corporate Change*, **9** (2), 211–53.

Godec, M., V. Kuuskraa and B. Kuck (2002), 'How US Gulf of Mexico development, finding, cost trends have evolved', *Oil and Gas Journal*, **100** (18), 52–60.

Hayward, A. B. (1999), 'Exploration frontiers for new century determined by technology, politics', *Oil and Gas Journal*, **97** (50), 42–4.

Langlois, R. N. (1992), 'Transaction cost economics in real time', *Industrial and Corporate Change*, **1** (1), 99–127.

Langlois, R. N. (2002), 'Modularity in technology and in organization', *Journal of Economic Behavior and Organization*, **49** (1), 19–37.

Langlois, Richard N. and Paul L. Robertson (1995), *Firms, Markets and Economic Change. A Dynamic Theory of Business Institutions*, London and New York: Routledge.

Leonard-Barton, D. (1992), 'Core competencies and core rigidities: paradox in managing new product development', *Strategic Management Journal*, **13**, 111–125.

Leonard-Barton, D. (1998), *Wellsprings of Knowledge: Building and Sustaining the Sources of Innovation*, paperback edition, Boston, MA: Harvard Business School Press.

Loasby, B. J., (1998), 'The organization of capabilities', *Journal of Economic Behavior and Organization*, **35** (2), 139–60.

Loasby, B. J. (1999), *Knowledge, Institutions and Evolution in Economics*, London and New York: Routledge.

London Infield Systems Ltd. (2002), 'The world deepwater report IV 2003–2007', accessed 12 January, 2003 at www.deepwater.co.uk/Deepwater_Report2003–2007.htm (December).

Marris, R. (1966), *The Economic Theory of Managerial Capitalism*, London: Macmillan.

Moritis, G. (2000), 'Industry confronts challenges', *Oil and Gas Journal*, **98** (18) 91–7.

National Ocean Industries Association (NOIA) (undated), *History of Offshore*, accessed 13 January 2003 at www.noia.org/info/history.asp.

Pavitt, K. (1998), 'Technologies, products and organization in the innovating firm: what Adam Smith tells us and what Joseph Schumpeter doesn't', *Industrial and Corporate Change*, **7** (3), 433–52.

Penrose, E. T. (1959), *The Theory of the Growth of the Firm*, Oxford: Blackwell.

Rhodes, A. (1998), 'Vastar uses technology, strategy to compete with majors in deepwater', *Oil and Gas Journal*, **96** (49), 27–33.

Richardson, G.B. (1953), 'Imperfect knowledge and economic efficiency', *Oxford Economic Papers*, **5**, 136–56.

Richardson, G. B. (1972), 'The organization of industry', *Economic Journal*, **82** (327), 883–96.

Richardson, G. B. (2002), 'The organization of industry re-visited', *Danish Research Unit for Industrial Dynamics working paper 2002–15*, Copenhagen.

Robertson, J. D. (1999), ARCO new ventures director, personal interview, Plano, TX, 17 November.

Rose, B. (1999), 'Consolidation, changing core competencies alter offshore drilling responsibilities', *Oil and Gas Journal*, **97** (50), 52–60.

Sanchez, Ron and Joseph T. Mahoney (1996), 'Modularity, flexibility and knowledge management in product and organizational design', *Strategic Management Journal*, **17** (winter special issue), 63–76.

Shirley, K. (2002, October), 'Global depths have great potential', accessed 11 January, 2003 at www.aapg.org/explorer/2002/10oct/appex_deepwater.html.

Thomas, M. and D. Hayes (1999), 'Delving deeper' *Deepwater Technology*, supplement to *Petroleum Engineer International*, **72** (5) 32–9.

US Geological Surveys (2004), 1997 play analysis, The cornerstone of the national oil and gas assessment, Energy Resource Surveys Program, accessed 6 July 2004 at www.energy.usgs.gov/factsheets/NOAGA/oilgas.html.

Watts, P. (1998), 'Vision and reality: realizing the commercial prospects for deepwater production', Presentation to the American Association of Petroleum Geologists International Conference, Rio de Janeiro.

Yielding, C. A., B. Y. Yilmaz, D. I. Rainey, G. E. Pfau, R. L. Boyce, W. A. Wendt, M. H. Judson, S. G. Peacock, S. D. Duppenbecker, A. K. Ray, R. Chen, R. Hollingsworth and M. J. Bowman (2002), 'The history of a new play: Crazy Horse discovery, deepwater Gulf of Mexico', Presentation to the Association of American Petroleum Geologists' Annual Meeting.

5. Consumers and suppliers as co-producers of technology and innovation in electronically mediated banking: the cases of Internet banking in Nordbanken and Société Générale

Staffan Hultén, Anna Nyberg and Lamia Chetioui

1. INTRODUCTION

After a very media-hyped start many start-up e-banks now face serious economic problems. In contrast the traditional banks using a multi-channel strategy are rapidly gaining market share in the Internet channel. Internet banking is the latest step in the development of distance banking. Internet and other distance banking services hold interesting properties for the banking firms and for the customers. The bank's principal benefits are increased customer retention and lower costs due to the transfer of work to the customers. The customers' most important benefits are increased accessibility of the bank service, lower service charges and avoiding having to wait to be helped by a teller.

The Internet is not only a new distribution channel for the bank sector but it also modifies the bank sector's competitive landscape. New competitors, for example specialized share traders on the Internet, specialized mortgage services on the Internet and Internet banks without bank branches, have attacked the banks by promising to provide a more rapid and cheaper service. Information and communication technologies (ICTs) have already resulted in two evolutionary change processes in the banking industry. In a first development stage they supported the processing of the bank's internal business and inter-bank transactions. In a second development stage they became the fastest channel to get access to the capital markets and permitted the creation of global electronic marketplaces. Today, ICTs provide support for the banks' branch network commercial operations and the development of new ways to organize distance banking services. Distance banking over electronic media appeared in the 1980s. Pennings and Harianto (1992) found one US bank that adopted videotext banking in 1981; in 1985, 37 out of 152 studied US banks had adopted videotext banking. The first commercial in-home banking system in the UK was launched in 1983 using a videotext system. The early systems using a micro-computer or other terminals linked by telephone or videotext had limited success due to too high costs and narrow services (Wright and Howcroft, 1995) or regulation (Pennings and Harianto, 1992). This was also true in France where the minitel bank, despite attracting many more customers than in the UK or the USA, reached only a few per cent of the private bank customers.[1]

It has been suggested that the French banks, because of their success with minitel, are lagging behind their European competitors in the adoption of Internet banking. The argument is that the French banks invested too much effort in transposing their minitel services to the Internet without investigating the possibilities of the Internet's higher interactivity levels than minitel. From a commercial point of view the business model of minitel was built on relatively high access charges which many French banks sought to impose on their Internet clients who were becoming accustomed to the Internet's spirit of cost free services.

In this chapter we will study how banks use external resources to develop an Internet bank. In particular, we will focus on differences in the role of suppliers of technology and the customer involvement in the transition to Internet banking. To do this we have chosen to study the adoption of an Internet bank in two different cases, namely in Nordbanken, the Swedish partner bank in the Nordic bank Nordea and in the French bank Société Générale. Both banks were renowned for their success with electronically mediated banking before Internet banking. Société Générale had both a telephone bank and a minitel bank. Merita, the Finnish partner bank in Nordea, had a telephone bank and PCs connected over the telephone network, and Nordbanken, the Swedish branch of Nordea, had a telephone bank. The two banks subsequently showed big differences in the transition to Internet banking. Prior studies of innovation in services have pointed to the role of external actors in shaping or even driving the development. In analysing and comparing these two case studies, therefore, we have paid particular attention to the role of such external resources – suppliers and customers of the bank. Empirical material for the case studies was collected through interviews with bank managers and through bank internal documents and other secondary sources. For a list of interviewees see Appendix Table A5.1.

The chapter is organized in the following way. Sections 2 and 3 present the theoretical concepts and models that we use to analyse and compare the two banks' development of Internet banking. Sections 4 and 5 present the two case studies. The chapter ends with a discussion and concluding remarks.

2. FIRST MOVER ADVANTAGES AND DISADVANTAGES AND INTERNET BANKING

The creation of an Internet bank is a major undertaking for most banks. The banks are confronted with a new competitive situation in which the old cost advantages and customer relations are changing. But there exist opportunities for incumbents in such turbulent markets. They can for example, use their historical market position to move to markets where the advantage continues to be significant.

> Companies slow to accept the inevitability that new technologies will force lower prices for basic information may find themselves losing market share rapidly on all fronts. Competitive advantages based on access to raw information are under siege; the trick is to migrate incumbency and scale advantages into value-added aspects of information, where advantage is more sustainable. (Shapiro and Varian, 1999)

This may sound easy, but in reality such shifts in strategy are cumbersome and difficult. To make such a move towards Internet banking a bank must commit itself to many substantial strategy changes:

1. It must decide on how to organize and govern/control the new channel in relation to existing parts.
2. It must allow the customers to participate more actively in the carrying out of the bank service.
3. It must learn to make new cost and revenue calculations.[2] An important element of this is that the bank must question historical investments in bank branches and other interfaces with customers. The weight of the investments in bank branches and the competencies of the bank personnel to manage service interactions face to face with clients should be compared with the completely new service delivery system of Internet banking: home banking and machine to machine interactions.
4. It also has to make decisions on the marketing of the Internet bank service, because the increased involvement of the customers in the banking services gives both a higher value added for the customers (faster and more certain service delivery) and lower marginal costs for

Table 5.1   The new cost–benefit equation of Internet banking

Factor	Producer	Client
Cost/price savings	Less personnel, fewer branches and less office space	Lower fees and lower interest rates
Value creation	'Image' benefits, higher client retention, and learning	Flexibility and more privacy

the bank. This new equation (see Table 5.1) opens avenues for entrants to attack the position of incumbents. We have a case in which an entrant could get an advantage both by a perceived benefit strategy or a low cost strategy (Besanko et al., 1999).

In short, the bank has to negotiate a transition from one technology to another under the threat that other banks will make this move faster and be successful in adopting the new technology thereby gaining a first mover advantage (Robinson and Fornell, 1985; Schmalensee, 1982). We can envisage different reasons why first mover issues are pertinent to the development of Internet banking. On the one hand we can imagine that a first mover gets substantial advantages. According to Lieberman and Montgomery (1988), first mover advantages arise from three primary sources: (1) technological leadership, advantages derive from learning or experience curve effects, where costs fall with cumulative output and success in patenting or R&D; (2) possession of scarce resources; and (3) market position, late entrants must invest extra resources to attract customers away from the first mover firm. Switching costs typically enforce the gains of the market share obtained early in the development of a new market.

On the other hand we can construct a case in which a first mover is disadvantaged. (1) Internet banking is the latest development in distance banking over electronic media. Advantages accrued in earlier stages of electronically mediated banking services are not necessary transferable to Internet banking due for example to the first mover's 'incumbent inertia'. (2) Internet banking as such is a first mover advantage that may be difficult to sustain if latecomers can benefit from the first mover's investments (Lieberman and Montgomery, 1988).

Timing is an important issue in innovation processes that depend on technological change in interrelated businesses. A bank would like to avoid the two potential errors of adopting too early or too late. In the first case the first mover selects a design that becomes unsuccessful, in the second case the laggard will face an uphill battle against the first mover who entered with a right design at the right time (Pennings and Harianto, 1992, pp. 44–5).

Table 5.2   E-transaction potential of Internet bank services

Service	Transactions per year and client
E-commerce	>100
Bank account balance	10–20
Payments of bills	10–20
Bank transfers	10–20
E-loans	2–3
Foreign payments	1–10
Internet equity trading	>10
Student loans	2–4
E-salary	12

Source: Hultén, Nyberg and Hammarkvist (2002).

Two further complications with first mover advantages in Internet banking are that the customers' usage of the new service is still under construction and that the average number of transactions per customer continues to be low. In Table 5.2 we have made a list of different services that a bank customer can perform over an Internet bank connection. Even an inveterate user of the Internet bank service seldom makes more than five transactions per month. The biggest volumes remain to be created – for example high volume equity trading and e-commerce.

3.   INTERNET BANKING AS A SERVICE INNOVATION

In services as in other industries, innovation draws on both firm internal and external resources (see Tether, 2002). Among studies of innovation, the view of the relative importance of external to internal resources in service innovation varies (see Table 5.3). External sources of innovation may be (interactions with) suppliers or customers, or from drawing on new or existing technologies in the environment in general. The degree of influence that the service firm is seen to have can also vary from being portrayed as a passive adopter of a finished innovation, to being seen as taking an active part in its development and/or adaptation. Pavitt (1984) portrays service innovation as driven primarily by supplier industries, with the service firms being seen as passive adopters of already developed innovations. Normann (1991) represents quite an opposite view, emphasizing both the role of the service firm in developing innovations, and the clients as important sources

Table 5.3   Sources of innovations in services

Aspect	Externally driven	Internally driven	Interaction-driven
Suppliers as sources of innovation or inspiration	Pavitt view **Société Générale**		
Customers/consumers as sources of innovation or inspiration		**Nordea**	Normann view

of external resources. In the empirical analysis we will see cases where external resources are important, and cases where internal resources are important in driving innovation.

Internet banking belongs to the rapidly developing group of self service technologies (SSTs) (see Table 5.4). They allow market space transactions in which no interpersonal interaction is required between buyer and seller. SSTs typically offer advantages to customers in that they help customers save time and money, avoid service personnel, and are accessible anytime and anywhere. The disadvantages with SSTs are related to technology failure, process failure often involving the translation of an electronic message to a physical delivery, poor design which results in confusion, and customer-driven failure, for example when a client forgets a PIN code to a bank card (Meuter et al., 2000).

The introduction of service innovations typically involves a reshuffling of tasks between providers and customers. Normann (1991) distinguishes between two types of innovations – relieving and enabling. *Relieving* innovations are those where the consumer is relieved of performing a certain task which can profitably be performed by a commercial provider. *Enabling* innovations, on the other hand, provide the consumers with the tools and knowledge to perform the task themselves. Internet banking is a good example of an enabling innovation. By giving the bank clients access to their own accounts and to many financial services, the clients could take on tasks for which they previously had needed the help of bank employees.

Enabling innovations inevitably involve consumers as co-producers, or *prosumers*, to use the term coined by Toffler (1980). Customer co-production is typically rationalized in terms of cost savings. Because the scope for productivity gains is relatively smaller in service industries than in manufacturing, it is argued that consumer involvement is an alternative way to lower the costs of service production. In the case of Internet banking, the cost savings come from reductions among teller personnel, and fewer bank branches.

Table 5.4   Categories and examples of self service technologies in use

Interface / Purpose	Telephone/interactive voice response	Online/Internet	Interactive kiosks	Video/CD
Customer service	Telephone banking Flight information Account information	Package tracking Account information	ATMs Hotel checkout	
Transactions	Telephone banking Prescription refills	Retail purchasing Financial transactions	Pay at the pump Hotel checkout Car rental	
Self-help	Information telephone lines	Internet information search Distance learning	Blood pressure machines Tourist information	Tax preparation Television/CD- based training

Source: Meuter et al. (2000).

Some of these savings are passed on to Internet bank customers in the form of lower fees and better interest terms. In addition, consumer value is created by the ability of Internet banks to offer 24-hour banking, a higher degree of privacy in banking, an increased sense of control and perhaps also positive feelings related to being a 'modern' person. Internet banking can create value for the bank by allowing it to offer new types of revenue generating services – e-commerce transactions – and more easily accessible information on how the customers conduct their bank business.

Providing Internet banking builds up a customer base; experienced users who, if satisfied with the service, can recruit and teach potential new users. Interaction with the first users, such as through problem solving, also enables the bank to learn and improve its services. According to Normann (1991) the productivity of the service client can be improved in four different ways: (1) the most important inducement is costs; (2) the client can be educated; (3) the client can be given different tools, and (4) by creating constellations of clients that are beneficial to the service delivery system.

Internet banking entails significant elements of customer involvement in the innovation process. Similar examples of customer involvement in innovation have become a trademark of the Internet industries and software industries (see McKelvey, 2001 and Thomke and von Hippel, 2002). In Normann's model (1991) of client participation the client can participate in six different ways in the service industry: (1) specification of the service; (2) pure co-production; (3) perform quality control; (4) development of the service; (5) marketing of the service, and (6) maintenance of ethos by providing employees with interesting experiences. Since the Internet bank is an SST, the provision of ethos plays an insignificant role, while the other ways of client participation are significant. Figure 5.1 depicts the interplay of customer and bank actions in driving the innovations process associated with the Internet bank service. With this perspective we have an adoption process in which the service innovation evolves from producer–consumer interaction (Rosa et al., 1999).

4. CASE STUDY 1: NORDBANKEN

4. 1 Distance and Internet Banking at Nordea's Swedish Bank

Nordea is the biggest bank in the Nordic region with ten million private customers, 1.1 million corporate customers, 2.7 million private e-clients, 1260 branch offices and 40 000 employees (Nordea, 2002b). Nordea is the result of the merger in 1997 of the Swedish bank Nordbanken and the Finnish bank Merita. In 2000, the resulting bank acquired Kreditkassen in

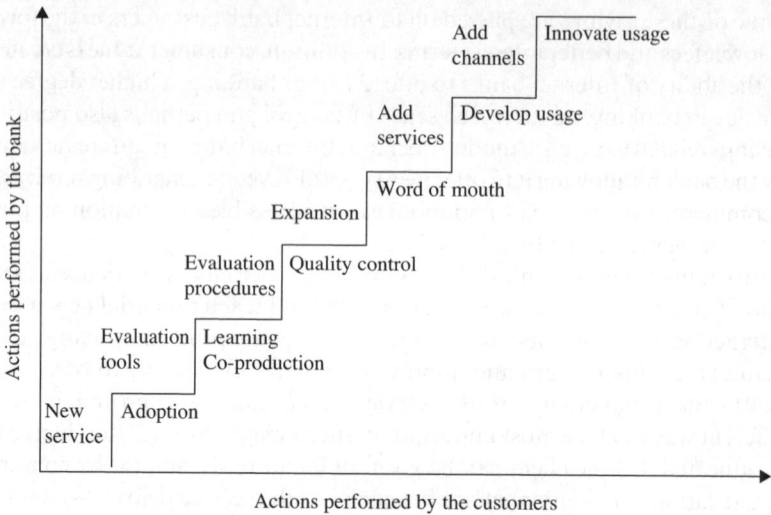

Figure 5.1   Co-construction of the Internet bank by the bank and its customers

Norway and Unibank in Denmark. Nordea also own banks in Estonia and Poland. This case will focus on the building of an Internet bank service at Nordbanken.

According to Nordea, it has the biggest electronic banking traffic, and the longest e-experience. Merita and Nordbanken combined had more than two million netbank customers in December 2001 – an increase of 25 per cent in 2001 from 2000. Unibank and Kreditkassen also offer Internet banking and the Estonian bank has a small Internet bank that uses the Internet bank model from Merita. Nordea is in the process of merging the different Internet models that have developed in the four different banks. In 2003 a common platform will be launched for the Danish, Finnish, Norwegian and Swedish bank operations in Nordea. The platform will first appear in Merita and then move to Nordbanken and the Danish and Norwegian partner banks (see Table 5.5).

4.2   Distance Banking over Electronic Media – the Telephone Bank at Nordbanken

Nordbanken emerged as the result of two merger steps in the 1980s. First Sundsvallsbanken merged with Upplandsbanken to become Nordbanken. This bank was merged a few years later with the former state-owned bank PK-banken to form a bigger Nordbanken.

Table 5.5   Main steps in the development of distance banking in Nordea

Year	Service
1982	IVR telephone service (Merita)
1984	PC bank (Merita)
1990	Full service telephone bank (Nordbanken)
1991	The brand Solo is introduced (Merita)
1992	Mobile telephone bank service (Merita)
1996	First Internet services (Merita and Nordbanken)
1997	IVR equity trading
1998	Improved web pages at Nordbanken
1998	Card solution at Norbanken
1998	Internet equity trading
1998	Introduction of CTI
1999	Short Message Service
1999	Wireless Application Protocol service
2000	Personal Digital Assistant service
2002	Chip services

Source: Nordea (2002a).

Distance or home banking at Nordbanken started in 1988 with a project aimed at a complete telephone bank service. Before 1988, Nordbanken offered different types of distance banking: for example, the possibility of paying bills through the Swedish banks' 'giro' system and a simple telephone bank where the customer could check their balance and the latest movements on the account. The origin of the telephone bank project was that many clients called their bank branches to ask questions. One problem with this was that the bank couldn't offer a full service since it was difficult to be certain about the identity of the caller and another problem was that calling customers were kept waiting because of too many incoming calls. The problem was particularly pressing in Stockholm where a bank employee was appointed project manager of a project to solve the problem. The project manager first designed a system, with the help of the Swedish telecommunication operator Televerket, that brought together six bank branches in central Stockholm to one switchboard.[3]

This solved the problem with the waiting time but the customers still could not get an attractive bank service over the telephone. During study trips to the USA two other managers at the bank had looked at American telephone banks. These two inputs resulted in a decision in 1988 to launch a project aimed at the creation of a complete telephone bank. The project was rapidly conducted under the direction of the project manager who had designed the system in Stockholm and a plan was put forward in 1989. The decision to launch the bank was difficult because the costs were considered to be high, in total 15–20 million SEK. The costs were due to investments in switches, marketing and a new office in Uppsala – 70 kilometres north of Stockholm.

The new telephone bank commenced operation in 1990. The technique was taken from the former PK-bank, while the bank personnel came from the former Upplandsbanken. This created some minor problems in the beginning when the personnel had to learn how the machines and the system worked. The project manager was appointed manager of the new telephone bank and he and the management of Nordbanken decided that the telephone bank should recruit a new type of bank personnel to the call centre. The manager of the telephone bank recalls: 'We sought social competence only.' This resulted in the bank employing a diverse workforce, for instance, employees who were nurses and bartenders. The relocation to the university town of Uppsala gave access to a large pool of university students who were willing to work part-time and who could come in on short notice to help cut peaks in the traffic.

Nordbanken became the first bank in Europe to offer a telephone bank that integrated an automatic service with a personal service. The telephone bank called 'Plus Direkt' was from the start organized as a separate bank within Nordbanken. The audiotel system made it possible to identify the customer. After calling the 'Plus Direkt' number the client inputs their personal number (all Swedes have a unique number consisting of their birth date and four extra digits) which is the cheque account number in Nordbanken.[4] Then the caller inputs a four digit PIN code. This gave access to some of the bank services, for example checking the balance on the bank account, and transferring money between their own accounts. If the customer wanted more advanced services, for example credit, or equity trading, dialling 91 connected the caller to a person at the call centre. Further identification consisted of providing a one time PIN code, which could be got from a list of codes that the bank had sent. These codes were printed on a paper with 20 PIN codes that were sent in a registered letter. After a code had been used, the customer barred it on the paper. This system was later changed to a system of cards with 50 numbers concealed under a thin film. The customer uncovered the codes stepwise.

The first year of the telephone bank was difficult. The bank branches were supposed to promote the telephone bank, but they were unwilling to do so because the transactions of telephone bank customers were not accounted for at the bank branch. The plan was that the telephone bank should have 30 000 clients after one year but, after eight months it only achieved one-tenth of the target. In the first months the office with 15 employees took 15–20 calls per day. To improve the enrolment of new clients it was decided that customers in the future should stay in the internal accounts as customers of the branch office and that the profit of the telephone bank activities should be transferred back to the respective branch office. The name of the telephone bank service was changed to 'Nordbanken Direkt'. The slow growth continued for a couple of years and the management at the telephone bank had to fight hard for the survival of the telephone bank. Nordbanken Direkt was introduced at a time when Sweden was in a severe financial crisis and Nordbanken faced big economic difficulties. However, the telephone bank received support from managers supervising the new activity, and accurate forecasts convinced top management that the telephone banking service was legitimate. It was not long afterwards that the new bank service started to grow faster than predicted.

The telephone bank service improved gradually during its first years. In the beginning, the call centre operator had to ask the client for his/her name and account number. The security check proved if the client was a telephone bank customer. At the start the work environment was not fully computerized – this was also the case in a traditional bank branch in the late 1980s. If a client wanted to buy or sell shares the call centre operator had to type the order on a typewriter. These type of problems created the insight that the telephone bank needed its own system to manage transactions and customers. When the bank developed such a system in 1992–93 it came to include items that we find in a market database or what we today call a customer relations management (CRM) database. The database included information on mailings to the customer and whether the customer had responded to these, the latest transactions, and the customer's overall connections to the bank. When this type of system was developed at Nordbanken little expertise was available outside the organization. Therefore Nordbanken had to develop a technique to connect a telephone operation to an in-house customer database. Competitive pressure from the Swedish bank rival SEB forced Nordbanken in 1992 to offer a 24-hour telephone bank service.

4.3 Internet Banking at Nordbanken

In 1995 the telephone bank manager in cooperation with the bank's information manager decided to move the telephone bank service to the Internet. They did this on their own initiative without any formal budget. They took some of the services of the telephone bank and created web pages for each option. The security level was not the highest possible. They kept the system with one time PIN codes on a piece of paper. Nordbanken asked the Swedish Bank Inspection about the regulations for Internet banking and received the answer that the Inspection had no objections since the Internet service only covered transfers between customers' own accounts.

The demand for an increased security level came from within the organization. It was the head of security who with the support of internal accounting that demanded that the bank switch to a system with card readers connected to the PC. When the client wanted to connect to the Internet bank he/she would swipe the card through the card reader. According to projections in early 1996 the card reader solution would become the industry standard within a couple of years. Contacts with Microsoft indicated that all PCs would be equipped with card readers by 1997–98 (interview with the Head of the Telephone and Internet Bank at Nordea Bank Sweden AB). In the meantime Nordbanken offered its Internet bank customers the option to install card readers on their home computers. In early 1996, Nordbanken had approximately 10–12 000 Internet customers. However, the card reader solution demanded three communication portals while a PC normally only has two communication portals. Despite these problems a high percentage of the Internet customers were willing to install the card reader. Because of the lack of standardization in PCs, software and browsers, Nordbanken had to create a support service that the customers could call and ask about the installation of the card reader. Many customers called and were advised on how to complete the installation. In many cases it was time consuming work that required the bank customer to have technical knowledge.

The launch of the card reader system coincided with the take-off of Internet banking in Sweden. Nordbanken gained more customers but lost market share in comparison with its competitors (see Table 5.6). The smart card solution delayed the bank for two years compared with the competing Swedish banks that opted for less advanced technical solutions.

Internet banking in Sweden began to develop in 1996. There were 202 000 Internet bank customers by the end of 1997 (Bankforeningen website). At that time nearly 529 000 Swedish households were connected to the Internet. SEB was the first bank to achieve a strong position in the market (Svensk Telemarknad, 2001). SEB used a calculator that generated numbers that the client typed when he or she wanted to get access to the Internet bank as a security system. The next bank to get a big installed base of Internet bank customers was Föreningssparbanken. This bank was formed through the merger of the centralized bank of the former savings banks and the farmers' cooperative bank. The bank has 4.1 million clients in Sweden. It also uses a system with a calculator that gave the number the client should use when connecting to the Internet bank. If a customer only wants to look at the balance of the account he/she can use a personal code. The third biggest bank, Svenska Handelsbanken (SHB), with 1.5 million clients in Sweden, offered

Table 5.6 Growth of Internet banking in Sweden 1996–2003

	1996	1997	1998	1999	2000	2001	2002	2003
SEB	–	91 000	230 000	270 000	540 000	630 000	690 000	720 000
Föreningssparbanken	–	65 000	170 000	290 000	850 000	1 100 000	1 350 000	1 410 000
SHB	–	10 000	100 000	150 000	320 000	349 000	530 000	620 000
Nordbanken	10 000	15 000	50 000	110 000	707 000	1 039 000	1 300 000	1 470 000
Small banks	13 700	27 000	84 000	163 000	418 000	600 000	630 000	880 000
No. of Internet bank accounts	33 700*	218 000	634 000	983 000	2 835 000	3 718 000	4 500 000	5 100 000
Nordbanken's market share	–	6.8%	7.9%	11.2%	24.9%	27.9%	28.9%	28.8%

Note: *the measures for the small banks are partly based on estimates.

Sources: Svenska bankföreningen, Nordea, annual reports of SEB, Svenska Handelsbanken and Föreningssparbanken for the year 2001.

its Internet service for free, but had relatively fewer Internet accounts. The other big banks charged less than 200 SEK per year. Nordbanken charged 8–16 SEK per month, SEB charged 10 SEK per month and 400 SEK per year for share trading. Föreningssparbanken charges from 155–180 SEK per year. According to an article in a Swedish paper the Internet banks had a comparable service level (Sparöversikt, 2002).

After the merger of Nordbanken and Merita in 1997, and even though the banks cooperated in the development of the Internet and distance bank, the Finnish bank remained more advanced. The solution to the problem with the card reader came from the merger with Merita. The Finnish bank also used a system with one time PIN codes but these codes had a higher level of security, as the codes came with seals. Nordbanken instead used cookies as a protection device. After discussions with the security organization at Nordbanken it was agreed that the solution with one time PIN codes gave a sufficiently high security level. On the other hand, it emerged that the Swedish bank had a better solution for the web pages and the presentation of the commercial offer and these were adopted by Merita (interviews with the Executive Vice President at Nordea and the Head of Private Retail Banking at Nordea Bank Sweden AB).

In March–April 1999 the new security system for Internet banking was launched at Nordbanken. According to the product manager of Internet services this resulted in a dramatic increase in the number of Internet clients. From 1996 to December 1998, Nordbanken, with the card reader system, had increased its Internet client base from approximately 10 000 to 38 000. Less than one year after the introduction of the one time code cards the bank had 330 300 Internet clients, in December 2000 the bank had 707 000 Internet clients and in December 2001, 1 039 000 Internet clients. In June 2003 the bank had 1 470 000 Internet clients. These figures can be compared with the total of 1.5–1.6 million active clients at Nordbanken. In 2003 the Swedish banks had a total of 5.1 million Internet bank accounts. 471 000 of these being company Internet accounts.

The first generation of Internet banks, derived from the telephone bank, offered only information regarding ones own balance, transfers between own accounts and payments over the 'Postgiro' and 'Bankgiro' systems. A year later in 1997, together with the smart card, Nordbanken presented a new improved Internet bank service. The Head of the Telephone and Internet Bank at Nordbanken pointed out that the web pages were better designed and the number of services increased. In addition to the earlier services they included loan applications, automatic control of the account numbers when payments were made, and the possibility of an automatic system for repeat payments. When the web pages were developed in 1997 a group at Nordbanken tested the designs and the manner in which to phrase questions on selected customers. The tests were carried out on papers representing the planned screens. The customers were asked if they understood the questions and what kind of follow-up questions they would like to have.

To the early version launched in 1997, additional services have been added. The product manager of Internet services lists the new services: more complicated loan applications like mortgages were introduced in 1998, in 1998–99 came the possibility to buy and sell shares in trust funds, and in 1999 the possibility to buy and sell shares. In 1999, an e-marketplace called Solo, originally developed by Merita, was attached to the Internet bank. The principal advantage for firms using Solo is that the bank can guarantee that the client has paid the bill before the purchase is finalized. In 2000 e-billing and e-salary were introduced. In 2003 Nordea in Sweden integrated the service of foreign payments that had been offered in Finland since 1999.

The client's access to the Internet bank costs from 8–16 SEK per month depending on the customer's relationship with the bank. Bank employees were given free Internet banking, which allowed the bank to rapidly increase their user base. The stronger the customer–bank relationship the lower the charge. The Internet service is able to combine with Internet share trading, and that service costs an additional 15 SEK per month. The charge for the basic Internet bank service is comparable to the charge the bank demands for a mailing service for paying bills through the centralized Swedish systems for paying bills. The mail-based systems normally handle a request for paying bills within two to three days depending on when in the month the letter with a signed payment order is posted. The systems handle many more bills at the end of the month because most salaries are paid on the 25th of the month and most bills should be paid before the 1st.

In December 2001 nearly 70 per cent of Nordbanken's active clients had an Internet account. In Table 5.7 we look at the Swedish picture. We take the sum of the number of Internet accounts and compare these with two measures: (1) the number of private Internet connections – this gives measure A in Table 5.7, and (2) the number of households in Sweden according to the Swedish statistical office's figure from 1999 – this gives measure B in Table 5.7.

Table 5.7 shows how Internet banking moved forward in two leaps. The first leap occurred in 1997 when 41 per cent of the households with an Internet connection had an Internet bank account. The next leap happened in 2000 when the number of Internet bank accounts tripled and actually surpassed the number of private Internet connections. The figure on Internet accounts in 2003 from Table 5.6 shows that Sweden had more private Internet bank accounts (4.7 million) than households in 2001 (4.3 million). There are two reasons why we can have more private Internet bank accounts

Table 5.7   Internet banking penetration in Sweden 1996–2001

Aspect	1996	1997	1998	1999	2000	2001
No. of Internet bank accounts	23 700*	218 000	634 000	983 000	2 835 000	3 718 000
No. of private internet connections	184 080	528 930	1 276 000	1 673 200	2 064 200	2 569 000
Internet banking penetration (A)	0.12	0.41	0.50	0.59	1.37	1.45
No. of households	4 349 000	4 349 000	4 349 000	4 349 000	4 349 000	4 349 000
Internet banking penetration (B)	0.005	0.05	0.15	0.23	0.65	0.85

Note: * this estimate is probably too low.

Sources: Table 3, Svensk Telemarknad (2001); SCB, Statistics Sweden for the measure of the number of households.

than private Internet connections and private households: more than one person in a household with an Internet connection has an Internet bank account; or employees may use their office Internet connection for their Internet bank affairs. Some persons can have more than one Internet bank account.

4.4   Summary of the Nordbanken Case

Experience gained from telephone banking provided Nordbanken with a number of transferable competencies that could be used in the Internet bank. The system with one time PIN codes was simple to use by the customers and easily transferable to the Internet bank. The customer relationship database gave knowledge about how to market new bank services that could be used in the Internet bank. The full service telephone banking indicated that clients were willing to conduct almost all types of bank services from home. The system with no or low charges in the telephone bank also proved to be attractive in the Internet bank.

The window of opportunity for Nordbanken's entry into the Internet banking business was longer than most players thought possible when the Internet service was developed. Nordbanken was an early starter, but lost that advantage with the too advanced security system. The bank lagged behind for two years (1998–99) before catching up in 2000. The figures in Table 5.3 suggest that few customers switched banks because of the availability or quality of an Internet bank service.

The customers were the origin of many novelties. First, it was the calls from customers who wanted information that triggered the idea to develop telephone banking. Second, customers' use of telephone banking resulted in an early development of a CRM system. Third, customers signalled that the card reader solution was too complicated by calling up and demanding assistance with the installation. The customers also provided negative feedback by not adopting the card reader solution. A sample of bank customers was also used when the web pages were being redesigned.

Suppliers seem to have played a less important role. Televerket helped develop switches for the telephone bank. The first development of the CRM system was made in-house. Microsoft and other computer industry firms' promise of the rapid arrival of new PCs led to the decision to use card readers.

5. CASE STUDY 2: SOCIÉTÉ GÉNÉRALE

5.1 Distance and Internet Banking at Société Générale

The goal in this case is to reconstruct the evolution of distance banking and Internet banking for private customers at Société Générale, a European bank group. Its main activities are retail banking, capital management, and investment and financial banking. The retail banking was reinforced by the takeover of Crédit du Nord. In total Société Générale has more than 12 million private clients and it is the leading retail bank in France, measured in turnover, number of bank branches (Société Générale and Crédit du Nord have together 2600) as well as the number of clients in France (7.5 million). Société Générale is also the bank that has the best image in France according to the polling firm IPSOS.

In 2000 the bank became the first French bank to offer six different channels to access the bank from a distance: (1) telephone; (2) PC via the Internet; (3) minitel; (4) account balance warnings over mobile phones; (5) WAP over mobile phones; and (6) interactive television. The number of customer contacts over distance banking channels increased by 34 per cent from 95 million in 2001 to 128 million in 2002. Some 35 million of these were logons by Société Générale's 675 000 Internet bank clients and nearly 41 million contacts were made by 1.6 million bank clients using the telephone bank.

5.2 Distance Banking over Electronic Media – the Creation of Minitel Banking

In the early 1980s France became the first country to launch a télétel system. The French system was connected to a Vidéotex network and it consisted of three principal components: the minitel terminal, an access network with servers, and France Télécom's system Kiosque Télétel for payments. All the French banks utilized this system to provide certain types of information to their clients. In 1984–85 Société Générale launched a test service that gave its clients the possibility to check the balance on their bank accounts. In 1987 the bank added the option to make transfers between the client's own bank accounts.

In 1989 Société Générale added a telephone bank service to the minitel service. The bank used France Télécom's system audiotel to provide a 24-hour and seven-day-a-week service for its customers. The audiotel included a kiosque audiotel which included in the user's tariff the cost of communication and a payment to get access to the service. In 1989 the telephone server was administrated by another firm. This firm played the role of a distributor that supplied the interface between the bank and the clients and the corresponding technologies. The bank transmitted to the other firm the information about the clients' accounts.

At the same time the number of minitel clients increased steadily and reached 120 000 by 1990. This created bottlenecks in the server and Société Générale decided to add new servers as the demand increased. Possibilities to expand the services were discussed. In 1993 a meeting of chief executives decided not to offer stock market transactions because it was considered risky and that the clients could make mistakes.

The abundance of minitel clients and the bricolage structure of the minitel product (duplication with nine servers with different access modes: 3615LOG1 . . .3615LOG9) resulted in a study on distance banking in the coming years. On 30 May 1994 a directors' meeting declared that the ambition of Société Générale was to be one of the leaders in terms of distance banking products and to meet the demands of its customers on this matter. According to a manager of distance banking services this was the time when people argued: 'distance banking is a revolution' and when there existed a willingness to create the service, and make the investments. One result of the renewed interest in distance banking was the decision to launch a stock market option on the minitel because the competitors offered it with success.

In September 1994 a project team were given the task of developing a distance banking platform for the private customers – this platform was called Banque à DisTance (BDT). The aims of the project were: (1) remake the ergonomy of the minitel; (2) add functions to the minitel, for example payments to external accounts, stock market transactions, and consultation of the stock market; (3) change access to vocal server; and (4) create a telephone platform.

The project group was rapidly moved to an organizational unit dedicated to distance banking directly connected to the group that managed the retail banking activity and organized customer relations. This gave it a relatively high degree of freedom. A manager of distance banking services explains: '. . . the banking world is a bit rigid, we had some sort of start-up spirit . . . We irritated everybody . . . Everybody was jealous of our position and the means we were given to work.'

In parallel a benchmark conducted by the information department resulted in a choice to use a new architecture called 'the technical platform BDT'. It would consist of a client–server architecture with accessibility 24 hours per day and seven days per week and the security system demanded by Société Générale's information system department.

5.3   Internet Banking at Société Générale

France Télécom opened its kiosque micro service in February 1995, a private transpac network that was accessible through the number 3601 and a PC with a modem. In 1995 there existed many different networks that enabled firms and private individuals to get connected to the Internet, for example e-world and Compuserve. In this competitive environment ' . . . some of ours [Société Générale's] competitors started moreover to utilize this new technology and already tested PCs . . . Société Générale had to position itself in the micro informatics offer . . . Our competitors advanced and it was necessary to maintain a dynamic brand image' (Société Générale, 1995).

In March 1995 the BDT team presented a document in which they stated that the kiosque micro offer was the most attractive choice because of its billing system, its openness towards other networks and its flexibility. In addition France Télécom promised that the system in the near future would accept downloading of software.

In November 1995 Société Générale decided to install a web server. At this stage most of the French banks had a web server, but no bank offered an Internet bank service. Some banks' websites showed their annual report while others also offered information about their services including answering questions from the clients.

To enhance its image in new technologies Société Générale decided to offer its clients a selection of information on the Internet. In December 1995, Société Générale bought a server and placed it with an intermediary to avoid all contact between the bank's information system and the server. The intermediary had the job of configuring the server, designing the web pages, and adapting them to an Internet environment. At the same time the BDT team suggested in a document that in the future, when the Internet will be used as a distance banking service, it should be integrated with Société Générale's system. But before that could happen all the security issues, for example coding, needed to be resolved.

In March 1996 the project to develop a platform ran into budgetary problems. This problem was resolved during a budgetary meeting at which priorities were set and a decision made on a more formalized budget.

In May 1996 the BDT team presented a document, with the functions that they thought were necessary to get an attractive PC banking service based on the kiosque micro. A project manager in charge recollected in an interview: '. . . we get a service with the same functionalities as minitel. It offers consultation, transfers, and stock transactions. We get it on PC, with an Internet technology.'

In July 1996 a report 'Information on Internet Products' was elaborated by the BDT team to realize the Internet project. Although the goals of the project were similar, the work on the Internet project was different from the kiosque micro over a private network. The report (Société Générale, 1996a) imagined a connection between the Internet project and the kiosque micro. ' . . . the web pages with information about the services for the Internet will be reutilized for the PC version of the minitel service because of the ergonomic proximity (both used PC as a means of connection) . . . Nevertheless, the form will be adapted to the Videotex technology . . . .' The information on the different services offered by the bank was planned to become identical. The Internet service appeared in January 1997 with general information accessible to everyone.

On 9 September 1996, a board of directors' meeting discussed the development of distance banking. The future of minitel compared with other existing services was discussed. It was considered probable that the minitel service, despite the efforts of France Télécom, would be confronted with intensified competition on the one hand, with call centres, and on the other hand, with PC connections in the near future. It was acknowledged that the work on the BDT platform was running late and consumed more financial resources than planned. Another problem concerned the delay relative to the competitors in offering a possibility to carry out stock market transactions over the minitel service. The board of directors suggested that the kiosque micro project should aim at avoiding more delays and that it should be coordinated closely with the Internet project. Société Générale, (1996b) also stated that: ' . . . it is important to be reactive, because when the security problems have been resolved, the project "bank on PC" may migrate to the Internet environment.'

In 1997 Société Générale investigated the options for distance banking with interactive television and mobile telephone. At this time Société Générale had started to receive emails from bank customers who wanted to know when an Internet bank service would be made available.

The kiosque micro project ran into more delays for different reasons. For example, it was difficult to find consultants who had expertise both in minitel and the Internet. Cost increases resulted in a budget overdraft of 1.4 million FF from the initial budget of 9 million FF. The costs were underestimated because the managers thought that it would be easy to construct web pages from the minitel environment. The project manager of distance banking using kiosque micro explained: ' . . . We had to remake everything. On the consultation we finally had to spend one year.'

Another cost was the connection kit to the Internet banking service – a CD-ROM with a manual. The connection kit was believed to solve three problems: a commercial problem, because it used the Société Générale's logo; a technical problem, because it used well known browsers; and a financial problem, because the kit reduced the need for a hotline.

A further problem was that France Télécom's service kiosque micro was delayed and didn't appear before the end of 1997. By that time the France Télécom system was adapted to Windows 98 which few PC users had installed. It so happened that Société Générale found out that competing banks offered Internet bank services and this was taken as a sign that the security problem was solved. In October 1997 it was decided that the Internet service should first be tested on clients living abroad before being offered to every customer at Société Générale.

Instead of launching the Internet or the kiosque micro Société Générale decided to improve its minitel service by offering the new BDT platform in September 1997. The most important changes were: (1) the computers supporting the connections were changed; (2) the access mode was changed with one access mode (3615 SG) replacing many different access modes (from 3615 LOG1 to 3615 LOG9); (3) the BDT platform was introduced which created a time difference between the customer's operations and the carrying out of the transaction by the bank; and (4) a possibility to subscribe online to minitel banking, and to look and research the movements on the bank accounts (Société Générale, 1997).

In January 1998, the project to start a PC banking service based on the kiosque micro system was abandoned due to three reasons: priority was given the Internet bank service, a too low estimated rentability, and lack of development of France Télécom's kiosque micro system. It was decided that the services targeted for the kiosque micro should be used in the future for the Internet service.

In May 1998, the first Internet services appeared at Société Générale in France. They included the consultation of bank account balances, the downloading of historical data to be used in a customer's budget planning software and the possibility to subscribe online to Internet banking. The Internet banking system was protected by a 40 bit secure sockets layer technology. At the start it was believed that the clients should pay for the Internet service. But due to different problems visible to the customers (the service was out of order or connections were extremely slow) it was decided to delay charging until the launch of the second version. For this version they intended to have an enhanced security system (128 bit) that in 1998 was new in the French market.

At this time Internet banking took off in France. During 1999, the supply of home banking witnessed a rapid acceleration with more than 90 banks offering Internet services. Some of the bigger banks in France, such as BNP-Paribas, Société Générale, Crédit Lyonnais, CCF, and CIC, are members of the Association Française des Banques (AFB).[5] These banks dominated the French Internet banking market in early 2000 (see Table 5.8).

On 4 June 1999, the second version with 128 bit coding was adopted that offered transfers between bank accounts and stock market transactions: this put the Internet service on a par with the minitel service. In 1998, before the launch of the second version the Internet service had 40 000 clients and the minitel had 208 000 clients. The new system benefited from more powerful computers and a new system architecture. Nevertheless, the first months of the new service were difficult. 'We changed the architecture, and we hoped that everything would turn back to normal. We put the system online, and

Table 5.8   Market shares in the French Internet banking market in 2000

Type of bank	Market share in Internet banking
Banques populaires	6.2 %
Crédit agricole	16.1 %
Caisses d'épargne	3.6 %
Crédit mutuel	14.9 %
Banques AFB	58.8 %

Source: AFB (2000).

it didn't work, and we had to wait two months to achieve an acceptable operation.' The project manager explains: ' . . . In November 1999, it started to become poor again. By December 1999, it was an appalling shame . . . but in February 2000, it worked like a marvel. To achieve a normal operation, it took two years.' Because of these problems the service continued to be offered free of charge. Officially in June 2000 Société Générale announced a permanently cost free Internet banking connection. However, the clients would still have to pay for many types of services, for example stock market operations, payment of bills, and automatic payments.

When the second Internet service arrived Société Générale had adopted a new strategy towards direct banking. A new division within the distribution division called SGdiffusion took charge of distance banking. It was decided that distance banking was to be regarded as a channel comparable to a bank branch. This change also meant that the BDT team lost its independence and its members were moved to different sections within the distribution division. Less than a year later the distant banking concept was abandoned and replaced by the notion of a multichannel structure. The plans to change the Internet bank to a synchronic system have step by step moved further into the future. A shift is now planned to take place in 2005 and the costs are estimated to be 250 million Euros.

The new multichannel structure facilitated the growth of the Internet banking service at Société Générale. In 2000 the number of Internet clients tripled and reached 260 000. In 2002, the number of Internet bank accounts at Société Générale reached 675 571. The subsidiary Crédit du Nord has approximately half as many Internet bank clients. Table 5.9 shows how Internet connections increased in France and at Société Générale and its most important competitor, BNP–Paribas, over the period 1996–2002. From the figures we can see that Internet banking penetration remains low in France. BNP–Paribas and Société Générale together had less than 1.5 million Internet bank accounts in 2002. This means that less than 10 per cent of their private clients have an Internet bank account. Also if we compare with the number of households (France had 23 810 000 households during 1996–2001 according to INSEE, the French statististical office) and private Internet connections (nine million in 2002) the number of Internet bank clients is relatively low.

Société Générale's competitors did not act vigorously to capture market share in the Internet bank market. BNP–Paribas had a six month advantage over Société Générale during 1999–2000, but lost this advantage in the following years. Pure-plays (i.e. Internet banks) like ING Direct and Egg entered the French market in 2000 and 2002. ING Direct had 320 000 Internet bank clients in September 2003 (El País, 2003) and Egg had only 125 000 French clients in the autumn of 2003. ING Direct claims that its French operation is profitable while Egg has financial problems.

Table 5.9   Internet connections in France and at Société Générale and BNP 1996–2002

Aspect	1996	1997	1998	1999	2000	2001	2002
No. of private internet connections	200 000	500 000	1 000 000	1 900 000	5 200 000	7 000 000	9 330 000
Société Générale Internet bank accounts	–	–	40 000	66 000	260 000	444 911	675 571
BNP Internet bank accounts	–	–	33 000	140 000	386 000	–	700 000

Sources: ART, www.invest.bnpparibas.com/fr/; Société Générale's documents and www.societegenerale.fr/, www.journaldunet.com.

5.4   Summary of the Société Générale Case

The experience from the minitel bank provided Société Générale with a number of transferable ideas that could be used in Internet banking. First, the minitel system introduced the idea of charging. Second, minitel had a high security level functioning on a special network (France Télécom's network) as opposed to the open Internet network.

The window of opportunity for Société Générale's entry to the Internet bank business was very long. The first mover French banks initially made small investments in an Internet service. The feedback from customers was not used extensively while the role of France Télécom had a big impact on technology choices. It was only when Société Générale noticed that competing banks had started Internet banking operations that the bank abandoned the France Télécom solution for Internet bank services. Financially strong pure-plays entered the market as second movers in 2001–02.

6.   DISCUSSION AND CONCLUDING REMARKS

When the Internet bank was introduced Nordbanken and Société Générale had connected between 3 and 5 per cent of their customers to distance banking. Nordbanken had only a telephone bank. Société Générale had a minitel and telephone bank. Between 1998 and 2002, both banks were able to capture significant market share in their Internet banking markets. The adoption of the multichannel strategy seems effectively to have blocked the advance of start-up Internet banks and traders. But important differences are evident if we look at the two banks. At the time of writing Nordbanken has already moved a significant part of its transactions to the Internet, while Société Générale is still working on moving its customer base to the Internet. Nordbanken is working on creating new markets and new channels while Société Générale still has to make the transition to synchronic Internet banking.

6.1 The Role of Customers in Developing Internet Banking

Let us first turn back to the Normann model of client participation in the service industry presented above. He suggested that the client can participate in six different ways in the service industry, five of which are relevant to SSTs. Though relevant, we will not discuss the marketing by word of mouth, as we do not have access to data on this factor. Let us now consider for the remaining four aspects, how the customers in the two banks behaved.

In the Nordea case the customers played an important role in the specification of the telephone bank service, as their frequent calls to the bank branches constituted a problem that triggered the launch of this service. However, in the Internet bank the customers played no comparable role.

Société Générale was not driven by customer actions in the launch of the minitel and Internet banking. As regards the Internet banking, the customers could potentially have played a role, as the bank received emails from customers asking about the Internet banking before it was launched. However, it is not clear that such customer reactions were taken into account in the bank's decision process.

6.1.1 Pure co-production
In both banks, co-production has increased as the banks have moved from the telephone and minitel banks to the Internet bank. The volume of co-production increases with the availability of more distance banking services, as each new enabling service also allows the customers to take on more of the tasks formerly performed by the bank. Co-production has also resulted in new services, permitting an increase in the total volume of services provided by the banks.

6.1.2 Performing quality control
The complaints about problems with the card reader at Nordea constitute a good example of how customers perform quality control of services.

Another example of how Nordea's customers participated in the quality control is when a selected group of customers tested the new web pages before the relaunch of the Internet bank service. In the Société Générale case, there is no clear example of customers performing quality control. One possible example of a negative signal could be the slow customer adoption of the asynchronic solution for e-banking transactions offered by Société Générale. The fact that we have seen relatively few examples of customers performing quality control in these two cases, may be an indication of the special demands that SSTs place on service providers, in detecting customer reactions to the service characteristics.

6.1.3   Development of the service
In the Nordea case, the questions asked by the customers when calling the early telephone bank, provided insights into the kind of services of a future telephone bank. This eventually resulted in a CRM system. Although Société Générale received emails from customers asking about the Internet bank before it was launched, Société Générale was not driven by such customer actions in the launch of the Internet bank.

In general, analysis of the behaviour of customers when using the Internet bank, provides insights into which services are found useful. In the Nordea case this is used to find ways to increase the number of monthly transactions per customer.

6.2   The Role of Suppliers in Developing Internet Banking

The suppliers of technology played a more important role in the first phase of distance banking. The Swedish telecommunication operator Televerket provided the equipment that brought together six Nordbanken bank branches in central Stockholm to one switch. France Télécom provided the minitel system for Société Générale's minitel bank. The system was connected to a Vidéotex network and it consisted of three principal components: the minitel terminal, an access network with servers, and France Télécom's system Kiosque Télétel for payments.

The technology suppliers were less successful in supporting the transition to Internet banking. Nordea's contacts with Microsoft indicated that all PCs should be equipped with card readers in 1997–98. This proved to be wrong. France Télécom promised Société Générale that the kiosque micro service launched in February 1995, a private transpac network that was accessible through the number 3601 and a PC with a modem, would in the near future accept downloading of software. Another supplier-related problem was that France Télécom's service kiosque micro was delayed and did not appear before the end of 1997. At that date the France Télécom system was adapted to Windows 98 which few PC users had installed on their PCs. This eventually prompted Société Générale to opt for a nonprivate network Internet banking service.

6.3   Differences in History and Strategy

It is apparent from the above comparison, that the two banks have differed considerably in their use of external resources, and that these differences help to explain differences in the outcome. In addition, differences in historical experiences and strategy of the two banks can further help to explain differences in choices made and market outcomes with regards to Internet banks.

We can note that the two banks had two different strategies when they adopted the Internet bank. Société Générale wanted to charge the customers because of the increased value added. This is logical in view of their past sucess with the minitel. Nordea, on the other hand, regarded the Internet service from a cost perspective and therefore decided to charge the customers a low fee or no fee at all. Nordea's strategy is based on the experiences of both Nordbanken and Merita. The cost focus of Nordbanken can be traced to their telephone bank first being started mainly as a way to divert costly balance questions away from the branch offices. As other services were added, the value creation for both bank and customers has become evident, but the cost focus has remained important. Together these backgrounds led Nordea to adopt a high growth, rapid penetration strategy which ultimately resulted in its present leadership position.

One question that has been the object of very much debate is the relative advantages of the so-called 'pure-play' retailers (Internet retailing only, for example Amazon) versus the 'clicks-and-mortar' (established retailers moving into Internet sales, for example Barnes & Noble). The debate concerns whether the strength in the new Internet technology of the pure-plays that were designed specifically to take advantage of the new set of opportunities, or the retailing experience of the established firms would provide the crucial advantages. At the height of the Internet boom, most the bets seemed to be put on the pure-plays, whereas today much of the evidence is rather in favour of the clicks-and-mortar firms. The cases of Nordbanken and Société Générale, both clicks-and-mortar operations, provide insights into what types of advantages are relevant when established banks enter into Internet business. As is shown in the case descriptions, pure-plays were not able to capture significant market shares as first movers. The cases also show that the previous experiences of the two banks differ considerably in terms of providing the sort of self service banking that forms a basis for Internet operations. Thus the existence of or a lack of previous banking experience in itself is not enough to predict the outcome of launching an Internet service. Instead, qualitative differences in the nature of existing self service banking services play an important role in shaping this outcome.

Table 5.10 summarizes some of the potential benefits of existing technology use for the introduction of Internet banking. Looking at this summary of potential benefits may give us an idea of the reasons for the differences in success of the two banks. Comparing the table with the case descriptions, we find that some of the potential benefits proved realizable while others turned out not to be adequate under the circumstances.

That the transferability of capabilities from the telephone banking can be beneficial is evident in the Nordbanken case. For example, the simple method of access code on a sheet of paper could be used in the early phase to get the Internet system up and running. Later on it was replaced with a system that had a higher security level, but in the meantime Nordbanken could launch the Internet bank and joint learning through customer use could start to take place. Similarly, building the Internet bank by transferring 'modules' from the telephone bank was a cost-efficient construction method but, more importantly, allowed Nordbanken to benefit from the value constituted by a customer base used to a similar design and pedagogy.

On the other hand, in the case of Société Générale, some potential benefits did not turn out to be actual advantages. For example, the bank's experience in procuring complex systems from an advanced developer was of little value when no such system was made available by the supplier. In fact, this experience may even have been detrimental in that it contributed to keeping Société Générale from starting to develop a system of its own. Neither did the potential advantage of having a customer base of users willing to pay for services prove to be of actual value in the circumstances. Although the prospect of services on the Internet starting to become fee-based has been discussed for several years, this is still a very marginal part of Internet business. Trying to launch a pay-for-use service slowed down the adoption of Internet banking in at least two ways. First, by not introducing a potential disadvantage of the existing minitel solution and thereby not giving minitel users this incentive to try the new technology. Second, the usage fee also constituted a disincentive for potential customers who had experience of other Internet services and who therefore had adopted the idea of the Internet as a medium that should be free of charge.

6.4   Concluding Remarks

In this comparison of the introduction of Internet banking at Nordea and Société Générale we have found that the two banks, despite pursuing very

Table 5.10 Factors supporting/complicating the shift to the Internet bank at Société Générale and Nordbanken

Factors	Acquired competencies with the minitel bank (SG)	Acquired competencies with the telephone bank (Nordbanken)
Factors supporting shift to Internet bank	• Internal knowledge of handling computerized customer self service banking • Customer base of users experienced in computerized self service banking and in the specific procedures used by the SG minitel bank • Experience in working with pay-for-use self service bank • Customer base of users prepared to pay for use of computerized self service banking	• Access procedures developed • Internal knowledge of how to create customer co-production system • Customer base of users experienced in self service banking and in the specific procedures used by the Nordea telephone bank • Experience that a separate unit blocked the enrolment of customers to the new channel
Factors complicating shift to Internet bank	• Experience in procuring complex self service system from developer (France Télécom) • Experience in working with pay-for-use self service bank • Customer base of users prepared to pay for use of computerized self service banking	• The relatively low security level pushed the bank to adopt a too high security level which was both costly and complicated for the customers

121

different strategies, end up with not very different relative market positions. Both banks belong to the group of the most Internet-advanced banks in their markets. In the language of Shapiro and Varian (1999) the banks migrated incumbency and scale advantages into value added aspects of information. However, in the case of Internet banking the clicks-and-mortar banks have benefited from a mix of scale and scope advantages. For example: (1) the possibility to mix face to face interaction with Internet banking; (2) the offer of a bundle of services to Internet clients that were already developed for the non-Internet market (e.g. share trading, funds, private loans); and (3) the ability to offer accessibility to the bank service over a multichannel bank service.

The big difference between the two banks' move into Internet banking is that Nordea in Sweden has moved a majority of its customer base to the Internet and that Société Générale only has moved 10 per cent of its French customers to Internet banking. An interesting question from the point of view of strategy is if this will matter for the long term market position of the banks. The slower adoption rate of Internet banking by Société Générale's customers may be strategically important for three reasons. First, if we compare the two banks' growth curves in the Internet market it is evident that Société Générale is following a different trajectory than Nordea. After five years of Internet banking in 2002, Société Générale had reached less than 10 per cent of its bank customers and acquired an installed base of nearly 700 000 Internet accounts. These figures are much lower than the comparative figures for Nordea in Sweden in 2001: more than one million Internet accounts and nearly 70 per cent of the active customers had an Internet bank account. Second, the French banking sector continues to be attacked by pure-plays that offer bank products that attract segments of the customer base. There exists a continued possibility that the pure-plays will become dominant in the Internet market. The Swedish bank market is more difficult to attack because nearly all bank customers have an Internet bank account with their main bank. The smaller banks and the new pure-plays in the Swedish market have not increased their market shares in the Internet bank market, but together they had nearly 900 000 Internet clients in 2003. Third, as we have shown in the case studies Internet banking can both decrease costs for the bank and enhance value for the customers. This means that a relatively lower Internet banking activity in a bank would give it higher costs and lower revenues in the medium to long run compared with a bank with a much higher Internet banking activity. Our conclusion is therefore that although first mover pure-plays in the French market were unsuccessful, banks like Société Générale may still be attacked by second mover pure-plays or foreign competitors.

Table A5.1   Schedule of personal interviews

Firm	Contact	Position in organization	Date of interview
Société Générale	Mr Benoît Lauret	Former manager of distance banking services	28 July 2000
Société Générale	Mr François Naphle	Project manager of distance banking using kiosque micro	27 July 2000
Nordea	Mr Bo Eriksson	Head of private retail banking at Nordea Bank Sweden AB, former telephone bank manager	7 March 2002
Nordea	Mr Kurt Gustavsson	Head of the telephone and Internet bank at Nordea Bank Sweden AB	25 January 2002
Nordea	Mr Bo Harald	Executive vice president, Nordea	6 March 2002
Nordea	Mr Anders Tholander	Product manager Internet services at Nordea Bank Sweden AB	24 May 2002

NOTES

1. Pennings and Harianto (1992, p. 44) regarded minitel as an 'astounding success'.
2. Tang (1988) discusses the problem of comparing the return on investment of an old technology protected by massive fixed investments and a new technology in which no investments have been made.
3. Televerket was the state-owned telecommunication operator monopolist. In 2004 Televerket had its name changed to TeliaSonera after the merger between Telia and Sonera.
4. The history of this is that the two Swedish state-owned banks, Pk-banken and the Post Office, after they started to cooperate, used the personal number as the client's cheque account number.

5. The French legal system identifies six types of banks: commercial banks, cooperative banks, savings banks, local public banks with social aims, financial institutions, and specialized financial institutions with the sole goal of lending money to a clientele.

REFERENCES

Association Française des Banques (AFB) (2000), 'Internet: quel constat pour les banques françaises?', July.
Autorité de Régulation des Télécommunications (ART) (2002), 'Internet, premier bilan'.
Besanko, D., D. Dranove and M. Shanley (1999), *Economics of Strategy*, 2nd edn, New York: Wiley.
El País (2003), 'ING Direct, "la bestia negra" de la banca tradicional', 22 November, p. 58.
Föreningssparbanken (2001), annual report.
Hultén, S., A. Nyberg and K.O. Hammarkvist (2002), 'Mobilising consumer competence to create technology transition paths – the example of the Nordea internet bank', paper presented at the Sixth International Conference on Competence-Based Management, Lausanne, 26–28 October.
Lieberman, M. B. and D. B. Montgomery (1988), 'First-mover Advantage', *Strategic Management Journal*, **9**, 41–58.
McKelvey, M. (2001), 'The economic dynamics of software: three competing business models exemplified through Microsoft, Netscape and Linux', *Economics of Innovation and New Technology*, **10** (2/3), 199–236.
Meuter, M. L., A. L. Ostrom, R. I. Roundtree and M. J. Bitner (2000), 'Self-service technologies: understanding customer satisfaction with technology-based service encounters', *Journal of Marketing*, **64** (3), 50–65.
Nordea (2002a), 'From e-banking to e-business', presentation by Bo Harald, 22 February.
Nordea (2002b), 'Will Money Talk?', presentation by Bo Harald, 7 March.
Normann, R. (1991), *Service Management: Strategy and Leadership in Service Business*, 2nd edn, New York: Wiley.
Pavitt, K. (1984), 'Sectoral patterns of technological change: towards a taxonomy and a theory', *Research Policy*, **13**, 343–73.
Pennings, J. M. and F. Harianto (1992), 'The diffusion of technological innovation in the commercial banking industry', *Strategic Management Journal*, **13** (1), 29–46.
Post-och Telestyrelsen, Den Svenska telemarknaden (2001), www.pts.se.
Robinson, W. T. and C. Fornell (1985), 'Sources of market pioneer advantages in consumer goods industries', *Journal of Marketing Research*, **22** (August), 305–17.
Rosa, J. A., J. F. Porac, J. Runser-Spanjol and M. S. Saxon (1999), 'Sociocognitive dynamics in a product market', *Journal of Marketing*, **63** (special issue), 64–77.
Schmalensee, R. (1982), 'Product differentiation advantages of pioneering brands', *The American Economic Review*, **72** (3), 349–64.
SEB, (2001), annual report.
Shapiro, C. and H. R. Varian (1999), *Information Rules: A Strategic Guide to the Network Economy*, Boston, MA: Harvard Business School Press.
Société Générale (1995), 'La banque sur micro-ordinateur', document de travail, March.

Société Générale (1996a), 'Information produit sur Internet', cahier des charges, July.

Société Générale (1996b), 'Projet banque à distance sur micro – clientèle des particuliers', dossier de synthèse, 25 October.

Société Générale (1997), 'Compte rendu comité projet BDT', 21 October.

Sparöversikt (2002), no. 4.

Svenskabankföreningen, www.bankforeningen.se/list.asp?Category_ID=statistik.

Svenska Handelsbanken, (2001), annual report.

Tang, M. J. (1988), 'An economic perspective on escalating commitment', *Strategic Management Journal*, **9** (special issue), 79–92.

Tether, B. (2002), 'The sources and aims of innovation in services: variety between and within sectors', Centre for Research on Innovation and Competition discussion paper no. 55, November.

Thomke, S. and E. von Hippel (2002), 'Customers as innovators: a new way to create value', *Harvard Business Review*, **80** (2), 74–81.

Toffler, A. (1980), *The Third Wave*, London: Collins.

Wright, M. and B. Howcroft (1995), 'Bank Marketing', in C. Ennew, T. Watkins and M. Wright (eds), *Marketing Financial Services*, 2nd edn, London: Butterworth Heinemann, pp. 212–35.

6. Technological shifts and industry reaction: shifts in fuel preference for the fuel cell vehicle in the automotive industry

Robert van den Hoed and Philip J. Vergragt

1. INTRODUCTION

In the past decades the automotive industry has allocated considerable amounts of resources to the development of cleaner propulsion technologies as alternatives to the internal combustion engine (ICE), such as the battery electric vehicle (BEV), the hybrid electric vehicle (HEV) and the fuel cell vehicle (FCV). Of these alternatives the FCV is generally seen as the most likely candidate due to its energy efficiency, low emissions and use of renewable fuel. Although the hype surrounding FCVs suggests a dominant design for FCVs exists, this is in fact not the case. Van den Hoed and Vergragt (2001) conclude that three technical differences between future FCVs can be discerned. First, whether or not the FCV is hybrid (where a FC-system is combined with a battery) or fully FC-based. Second, whether the FC is used to propel the vehicle, or is used as an alternative to the current accumulator/battery; in this configuration the ICE remains, but the 'more problematic' battery is replaced. Third, the preferred fuel for the FCV is still disputed within the industry.

With respect to fuel, ideally a FCV uses pure hydrogen to generate electricity. Given the problems related to hydrogen storage and the high costs of developing a hydrogen infrastructure, car manufacturers are actively studying alternatives such as *methanol* and a *clean hydrocarbon*. Methanol can be made of fossil fuels (like natural gas) as well as from renewables (such as biomass). The 'clean hydrocarbon' is in fact a reformulated gasoline; reformulated in the sense that the gasoline is 'designed' for specific use and requirements in FCVs. Unlike methanol, the reformulated gasoline is a nonrenewable fuel. For reasons of clarity the word *gasoline* will be used instead of 'clean hydrocarbon' in the remainder of the chapter. For practical use in a FCV, a reformer is required to reform methanol or gasoline into a hydrogen-rich gas. The hydrogen-rich gas in turn is directed into the fuel cell, which chemically turns hydrogen into electricity, which then propels the car via an electric motor.

Since 1996 fuel cells have been increasingly considered a serious option for car applications by the automotive industry. And with this increasing interest, the issue of fuel preference became more pertinent. Different consortia of carmakers combined with oil companies are in competition to set the dominant design with respect to fuel preference, much like the competition between VHS and Betamax in the electronics industry in the early 1980s. After a period of relative consensus in terms of preference, in recent years differences in fuel preference among car manufacturers can be observed. Furthermore, in a period of five to seven years a shift from methanol towards gasoline can be discerned. This shift has all kinds of infrastructural, economic, and environmental consequences. With respect to the environment for instance, the environmental benefits of a gasoline-fuelled FCV over an ICE vehicle are small if not negative (Hart and Bauen, 1998; Höhlein, 1998).

The theme of fuel preference for FCVs by the automotive industry leads to several intriguing questions. Which mechanisms are at play in the adoption or rejection of a certain fuel, both on an industry level as well as on an organizational level? Are choices more externally driven (conformity to external pressures) or more internally, interest-driven? What are the consequences of technological shifts, and how might policymakers try to influence the technology decision process?

This chapter attempts to tackle these questions and is structured in the following way. Section 2 introduces a model for technology adoption by firms. Section 3 discusses the consequences of fuel preference for the technical configuration of FCVs, environmental performance and technical performance, together with an historical overview of shifts in fuel preference on an industry level. Section 4 forms the analysis of the case, where mechanisms of change are discussed; given the leading role of DaimlerBenz/Chrysler and General Motors (GM) their motives for fuel preference will be examined.[1] Section 5 concludes.

Data are gathered using a combination of literature research, analysis of press statements and car company year reports, and interviews at five automotive companies active in FC technology development. Historical data were collected via the Business and Industry database, *Financial Times* database, and the Fuelcells.org monthly newsletter. In order to substantiate the press releases with more in-depth knowledge of the decisionmaking process within automotive companies, 20 interviews were held with senior company representatives in positions either related to the R&D department (responsible for the FCV development) or the department responsible for environmental issues (see Table A6.1).

Apart from a qualitative assessment of fuel preference through interviews and press releases, a more quantitative approach was used by analysing fuel use in demonstration vehicles per firm. Although having some drawbacks (secrecy issues of demonstration models, PR function of models), the technical systems presented in demonstration vehicles do give an indication of R&D activities and allocation decisions within firms, and are considered a relatively good indicator of a firm's fuel preference.

2.   MODELLING R&D DECISIONS OF FIRMS

In order to understand technical decisions made by firms, R&D decisions by firms are modelled by using institutional theory (DiMaggio and Powell, 1983; Hoffman, 2000; Scott, 1995), and technology dynamics (Dosi, 1982; Nelson and Winter, 1982). The dependent variable in this model (Figure 6.1) is the decision whether firms will either conform to or deviate from established (technological) routines. Three elements of the model will be discussed: (1) the institutional embeddedness of firms; (2) institutional change through the organizational field; and (3) firm specificity.

2.1   Institutional Embeddedness of Organizations

Two assumptions underlie the choice for institutional theory. First, it is assumed that the automotive industry can be characterized as highly regulated, and highly uncertain with respect to future technological developments (e.g. FC technology, new technologies), regulatory standards (e.g. emissions, energy efficiency), and economic developments (e.g. oil price). Uncertainty can be defined as 'the degree to which future states of the world cannot be anticipated and accurately predicted' (Pfeffer and Salancik, 1978, p. 67). Although the current ICE is increasingly under pressure, it is highly uncertain whether and when technological shifts might occur, and which technological shifts are most likely to occur. The consequences of firms' actions are not calculable, and thus firms are more dependent on routines, standards and 'gut feeling' rather than on traditional risk analyses and calculation models (Levy and Rosenberg, 2002).

Second, it is assumed that technology is not an objective given, but is constructed by actors in the field. Competing technologies are negotiated in a broader organizational context, in which competing firms with competing technologies will try to get support for their specific technology in order to serve their interests. This can partly explain why 'the best technology' does not always win; in some cases a suboptimal technology might win due to the support of key actors in the industry. Technology choices are thus not only based on technological merits, but also on a political negotiation process between relevant actors.

Institutional theory is useful for describing cases in which uncertainty plays a large role, and in strongly regulated industries like the automotive industry. Central to institutional theory is that organizations are embedded in an external (institutional) environment (Scott, 1995). It stresses that organizational behaviour is to a large extent influenced and shaped by this institutional environment, which takes the shape of rules, laws, routines, values and traditions. These rules and routines prescribe what is to be seen as appropriate or legitimate behaviour for organizations. Not conforming results in penalties by powerful enforcers; as a result organizations are likely to conform to the institutional pressures. They do so in order to gain legitimacy. Organizational behaviour is then not always the most efficient or profit maximizing: firms might well choose activities which are not the most efficient but rather conform to the institutional pressures. Similarly firms might well remain in taken for granted routines, even though changing them would produce more efficient returns (Oliver, 1997).

Three aspects (or pillars, Scott, 1995) of institutions can be distinguished: coercive, mimetic and normative (DiMaggio and Powell, 1983). The *coercive* (or regulative) aspect of institutions relates to the formal and informal pressure exerted on organizations upon which they are dependent. Coercive pressures take the form of enforced rules, standards or mandates. Typical examples of coercive pressures include mandates or standards set by governmental agencies. But also powerful actors in the firm's environment can impose standards, for instance powerful firms mandating suppliers to acquire certain quality standards.

Mimetic pressures result from efficient response to uncertainty. When faced with uncertainty it can be economic for an organization to mimic best practice in order to reduce search costs. It serves as a convenient source of practices, which may be copied unintentionally. Mimetic isomorphism is often associated with a contagion process 'spreading fashionable features of one organization to another' (Haveman, 1993). Japanese modernization, and their copying of Western practices are mentioned as examples. Furthermore the copying of profitable new markets in the auto industry such as the SUV (Sports Utility Vehicle) segment serves as an example.

Normative pressures relate to the norms, standards, convictions, appropriate codes of conduct within an industry. They stem from *professionalization*: the struggle of an industry to define the conditions and methods of their work, and to develop a common language and create a form of collective rationality. Although institutional theory does not make specific mention of the role of technology in the institutional environment, the merits, bottlenecks, convictions and expectations surrounding a particular technology can be seen as normative aspects in the institutional environment. To illustrate, the dominance of the ICE in the automotive industry is surrounded by the widespread industry conviction that this is the best technology to do the job of propelling cars. But also discussions on the so-called 'platform strategy' or JIT (just in time) manufacturing principles illustrate widespread convictions within the industry, to which the industry as a whole conforms.[2]

Technologies tend to develop along technological regimes or technological trajectories (Dosi, 1982; Nelson and Winter, 1982). These regimes prescribe rules and directions of technological decisions, but also provide boundaries to technological directions. Green et al. (1994, p. 1056) note:

> The (technological) regime will ... constitute a set of 'socially' agreed objectives, as to what the parameters of an industry's products will be, how they would be typically made and, crucially for R&D, on which features of the product and process technological development should focus: in other words, which performance characteristics will serve as a heuristic for R&D attention.

Technologies are thus as much part of the institutional environment as are for instance regulations, and they manifest themselves in the form of convictions, norms, standards, and search heuristics for R&D to which firms will be likely to conform.

These three aspects of institutions influence organizational behaviour, both in a constraining way as well as in an empowering way (by providing notions of what is appropriate and how a firm should behave in certain situations) (Hoffman, 2000, Scott, 1995). Given similar institutional environments, firms will tend to look alike, leading to isomorphism (DiMaggio and Powell, 1983). Although the institutional forces constrain organizational activities, not all organizational behaviour can be seen as pure conformance behaviour: firms are thought to have room to manoeuvre within the limits set by institutional pressures (Greenwood and Hinings, 1996; Hoffman, 2000). Within these limits firms can either remain within the established routines, or deviate from them.

2.2 Institutional Change through the Organizational Field

Institutional theory traditionally connotes stability, but in recent years the process of institutionalization (Hoffman, 2000; Scott, 1995) or deinstitutionalization (Oliver, 1992) has received more attention, which permits us to study (technological) change within industries. Hoffman (2000) argues that institutions are changed through a negotiation process between relevant actors in the *organizational field*, defined as: 'a community of organizations (or actors) that partakes of a common meaning system and whose participants interact more frequently and fatefully with one another than with actors outside the field' (Scott, 1995, p. 56). The organizational field may include regulatory agencies, interest and pressure groups, consumers, and other public or private actors.

The organizational field is the centre for dialogue and discussion between relevant actors. It is at the level of the organizational field that meaning is developed with regard to issues arising in the field, in a process of negotiation and discourse between actors in the field. The organizational field then becomes an 'arena of power relations' (Brint and Karabel, 1991), and forms the locus where institutions are changed and developed. This change manifests itself in regulative, normative and cognitive aspects of the institutions as proposed by Scott (1995). To illustrate, the organizational field regarding sustainable mobility comprises actors such as car manufacturers, oil industry, regulators, and pressure groups. Through a process of dialogue new standards are set by regulators (regulative pressure), best practices are put forward (mimetic pressures) and standards and norms are developed concerning appropriate and legitimate behaviour (normative pressures).

2.3 Firm-specificity and Institutional Entrepreneurship

A last element of the model focuses on the level of the organization itself. Institutional theory underemphasizes a firm's history and culture as a determinant of its behaviour. Firms have specific backgrounds, traditions, beliefs and values (Nelson, 1991). They will influence the selection of appropriate activities within the spectrum of the firm-internal culture. In order to explain deviation strategies it can be expected that firm-internal aspects will be dominant, as institutional pressures tend to lead to conformance.

Three factors are used to distinguish between firms: resource position, power distribution, and internal routines and beliefs (Giddens, 1984; Nelson, 1991). First, resources include both the financial and nonfinancial (capabilities, competencies, external network ties). Resource serves both as an enabling factor for adopting new behaviour, as well as providing direction in selecting which activities best fit with their routines and capabilities, while preserving the current power situation.

Second, *power distribution* indicates which departments in the company are more capable in allocating resources to serve their interests. Firms with a stronger R&D department are more likely to get funds to carry out new technological research. Similarly, a more marketing-oriented firm will be more likely to focus resources on market introduction, consumer research and testing.

Third, firm-internal *routines and beliefs* are the more intangible aspects of firms, including both conscious ways of behaviour, as well as more taken for granted rules. Each firm will have their beliefs about how to be successful in the market, how to develop a good product, and what drives success for their organization. Certain firms might be aware of their strength in R&D; others may boast their strategy as a follower of others, while again others might mention their market knowledge as their key success factor. Internal routines are historically built (Nelson and Winter, 1982), and will influence technological decisions within firms.

2.4   Conceptual Model

Figure 6.1 shows the general model used for this study to understand R&D decisions of firms, and more specifically whether a firm is more likely to deviate from or conform to the established technological regime. It is important to distinguish three levels. At the level of the organizational field relevant actors negotiate over meaning, resulting in new standards and regulations (coercive pressure), best practices (mimetic pressure) and new rules and codes of conduct (normative pressure). Second, the institutions influence and constrain organizational behaviour, leading firms to conform

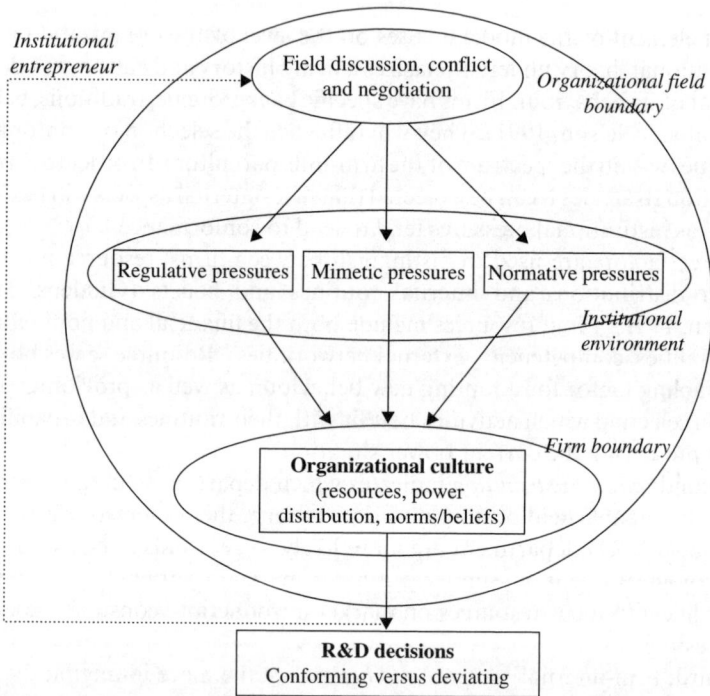

Figure 6.1   A model of R&D decisionmaking within organizations

and look alike. Third, the organizations themselves differ historically (resulting in different resource positions, power, routines and beliefs), which in turn make some firms more likely than others to conform or to deviate from the established (technological) regime.

3. FUEL PREFERENCE IN THE AUTOMOTIVE INDUSTRY

3.1 The Rize of Fuel Cell Vehicles

In 1990 the California Air Resources Board (CARB) issued the Zero Emission Vehicles standard (ZEV), mandating car companies to sell at least 2 per cent of their cars with zero emissions in 1998, increasing to 10 per cent in 2003. As the ICE was not able to achieve such standards, automotive firms started studying alternatives that could. Large investments were made in the development of the Battery Electric Vehicle (BEV); the FCV was not yet considered until the middle of the 1990s; at that time BEV technology was seen as a more realistic alternative to the ICE.

The first car manufacturer to actively start researching FCVs was DaimlerBenz by demonstrating the Necar1 in May 1994 (Kalhammer et al., 1998). It was only after the demonstration of the Necar2 in May 1996 that close competitors of Daimler began to see opportunities in FC technology for automotive use. By then the space requirements of the FC system had decreased by several factors, and the performance characteristics of fuel cell technology showed spectacular progress (Kalhammer et al., 1998). In January 1997 DaimlerBenz set up a structured program in the so-called Dbb Fuel Cell House, with the specific task to develop all FC components to the point of mass manufacturability.[3] In April 1997 Daimler bought a 25 per cent stake in Canadian Ballard (a $400m contract), the leading FC manufacturer globally.[4]

Daimler's activities and relative success with the Necar1 and Necar2 sensitized the rest of the automotive industry to also set up FC programs. Given that most companies in that period were focused on bringing a commercial BEV to the market these programs were relatively modest in comparison to Daimler's. Only in late 1997, when most BEV programs of ZEV-affected companies led to the market introduction of BEVs in 1998, automotive firms shifted their programs from BEV to FCV. Since this period all major car manufacturers have set up FC programs investing $25m–200m yearly to bring FCVs to a commercial stage. Kalhammer et al. (1998) report that the automotive industry had invested $1.5–2bn on the development of FCVs by the end of 1998. Programs have since become more structural and more market-driven, with yearly estimated investments[5] of the FC leaders DaimlerChrysler, General Motors and Toyota of $200–300m, providing employment for 500–600 people in the FC program.[6]

In the last five years FC technology has become the most prominent alternative to ICE for reducing emission levels, increasing energy efficiency in vehicles and paving the way to sustainable energy use (hydrogen). In the meantime BEVs have not been a success on the market, and most BEV programs in automotive companies were terminated around 1999. Van den Hoed and Bovee (2001) conclude that the ZEV regulation in California played a dominant role in providing a climate to change the current propulsion technology. It is further concluded that dominance of BEV was succeeded by dominance of FCV, and that mimicking behaviour between automotive companies was a dominant mechanism in describing the industry reaction to FC technology.

3.2 Fuel Options for FCVs and their Consequences

Figure 6.2 provides an overview of the three dominant fuels mentioned by automotive companies in relation to FCVs. On the right side of the figure the consequences of the shift from (1) hydrogen to methanol and (2) methanol to gasoline are mentioned on several parameters: environmental, infrastructural, technical, industry stakeholders, commercialization potential and consumers. The most important consequences are discussed below.

Technically the simplest configuration for a FCV is the one using hydrogen as the main fuel. FC systems require pure hydrogen in order to function efficiently, and should not be negatively affected by impurities such as carbon monoxide (CO). The main advantage of such a hydrogen-based system is the fact that no reformer technology is required (to reform another fuel into a hydrogen-rich gas), and that hydrogen is potentially a sustainable energy resource (when produced for instance via electrolysis of water by photovoltaics). The main barriers to this system are infrastructural (lack of hydrogen infrastructure, and high costs of setting up an infrastructure for gaseous hydrogen), technical (hydrogen storage remains costly and a heavy technical solution), as well as consumer-related: hydrogen is still associated with its explosive nature, and refuelling one's car with gaseous hydrogen will require a different routine for consumers.

In order to overcome the major technical and infrastructural barriers, methanol was proposed as an intermediate fuel (first by DaimlerBenz). Methanol requires the instalment of a so-called methanol reformer, which reforms methanol into a hydrogen-rich gas, which in turn fuels the FC system. Given the sensitivity of the FC system to impurities, the reformer has severe limits on its impurity level, which forms a major technical challenge.

Consequences of shifts: 1 Hydrogen → Methanol   2 Hydrogen → Gasoline

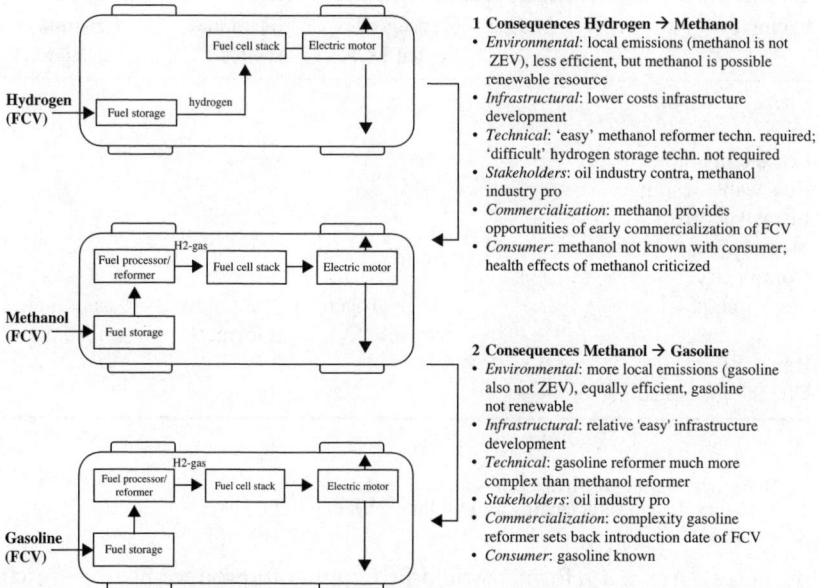

1 Consequences Hydrogen → Methanol
- *Environmental*: local emissions (methanol is not ZEV), less efficient, but methanol is possible renewable resource
- *Infrastructural*: lower costs infrastructure development
- *Technical*: 'easy' methanol reformer techn. required; 'difficult' hydrogen storage techn. not required
- *Stakeholders*: oil industry contra, methanol industry pro
- *Commercialization*: methanol provides opportunities of early commercialization of FCV
- *Consumer*: methanol not known with consumer; health effects of methanol criticized

2 Consequences Methanol → Gasoline
- *Environmental*: more local emissions (gasoline also not ZEV), equally efficient, gasoline not renewable
- *Infrastructural*: relative 'easy' infrastructure development
- *Technical*: gasoline reformer much more complex than methanol reformer
- *Stakeholders*: oil industry pro
- *Commercialization*: complexity gasoline reformer sets back introduction date of FCV
- *Consumer*: gasoline known

Figure 6.2   Technical designs of differently fuelled FCVs and consequences of shifts in fuel choice

Nevertheless, methanol reforming technology is generally seen as closer to commercialization than hydrogen storage technology.

Basic advantages of methanol over hydrogen are the relatively easy infrastructure development, due to the fluid nature of methanol. Furthermore, methanol has a higher energy density than hydrogen that positively influences the range of the methanol FCV. Also methanol can be produced from biomass, keeping the sustainable nature of the FCV intact. Lastly there are few consequences for consumers. Most critical barriers for the methanol FCV include energy efficiency loss of the system (reformers have an efficiency of 60–80 per cent), and the existence of (low) emissions of carbon monoxide (CO), carbon dioxide (CO_2) and NO_x; in other words, the main reason for FCV (zero emissions) is not achieved. In recent years the toxic nature of methanol has been emphasized by gasoline supporters. Another crucial barrier is that the oil industry has dismissed methanol: the costs of using methanol as a transition fuel towards the hydrogen society are too high, so it is argued.

The third and last fuel proposed by the automotive industry is gasoline (as said earlier in reality a clean hydrocarbon, reformulated for specific FC

Table 6.1   Comparison between different vehicle types

Factor	ICE	Hydrogen fuelled FCV	Methanol fuelled FCV	Gasoline fuelled FCV
Energy efficiency (well to wheel)	−	+[1]	+/−	−?[2]
Local emissions	−	+ +	+/−	+/−
Renewable resources	+/−	+	+	−
Infrastructure development	+ +	− −	+/−	+
Complexity technology	+ +	− −(hydrogen storage)	+/− (methanol reformer)	− (gasoline reformer)
Range of vehicle	+ +	−	+/−	+/−
Change for consumer	+ +	−	+/−	+

Notes:
1   Dependent on the way hydrogen is made.
2   Efficiency data are still lacking for gasoline reforming technology.

purposes). Again a reformer would be required to produce hydrogen-rich gas in the car for the FC system; however the gasoline reformer is much more complex to develop, and would delay the introduction of FCVs by years (Kalhammer et al., 1998). Also the environmental benefits are largely lost, given remaining emissions, lower efficiency (due to inefficient reformer), and continued use of fossil fuels. Advocates of gasoline, including the powerful oil industry, argue that neither infrastructural nor consumer changes are required. Table 6.1 summarizes some environmental and technical consequences of the different systems.

3.3   Industry's Shifting Fuel Preference for FCV

In assessing the shift in fuel preference for FCVs three periods can be distinguished in the 12 years to 2002: (1) 1990–1996, (2) 1997–1999, and (3) 2000–2002.

3.3.1   Period 1: 1990–1996
In the period 1990–1996 FC technology was still very much in the research stages, and experimental. The commercialization prospects were low; technical performance and the development of working models of fuel cell systems were of principal priority. Pure hydrogen was used in order to assess whether FCVs would indeed work, be drivable, and provide future

opportunities for improvement. Infrastructure issues were hardly discussed in this period, and technically the simplest fuel was chosen: hydrogen.

3.3.2 Period 2: 1997–1999

Due to spectacular progress, Daimler decided to bring its FC activities out of the research labs, into a development program at the end of 1996. Daimler expected that within three to four years the viability of FCVs could be evaluated, and a go/no go decision was mentioned for the year 2000. Given the addition of commercial goals to FCVs, it became necessary to think about the future fuel. Due to the problems of hydrogen storage and lack of infrastructure, there was general consensus in the automotive industry that hydrogen-fuelled FCVs would not be a viable alternative in the short term.[7]

Daimler was first to propose methanol as an alternative to hydrogen, around 1996–1997, and started investing heavily in the development of methanol-reformers.[8] In parallel it continued some activities in hydrogen storage and handling, despite its bad experiences with hydrogen storage technology in the early 1980s. Daimler's arguments for methanol included the relatively cheap set-up of a methanol infrastructure (in comparison to hydrogen), its easy reforming (in comparison to gasoline), the general availability of methanol (from natural gas) but also the potential renewability of methanol.[9]

Daimler's focus on methanol was part of a dual-fuel strategy. FC technology was considered for both the consumer market (passenger cars) as well as for the professional market (fleets, buses). The former (passenger cars) would require an extensive infrastructure of 'easy' fuel (i.e. liquid), for which Daimler suggested methanol. The latter (fleet cars, buses) would use central refuelling stations: for this market hydrogen made more sense.

By proposing methanol Daimler rejected the gasoline option, despite its 18-month partnership with Shell. Shell, not in favour of methanol, made an attempt to convince Daimler of the advantages and opportunities of gasoline reforming, using Shell's CPO (catalytic partial oxidation) process. Daimler however found the technology too complex, and was afraid that the introduction of FCVs on the market would be needlessly postponed by switching to gasoline. Shell and Daimler therefore terminated this partnership.[10]

What did Daimler's competitors do? At the time (1996–1998) most of the top ten car companies were investing considerably in BEVs as a response to the Californian ZEV regulation.[11] In 1997–1998 competitors like GM, Ford, Chrysler, Toyota, Honda, Mazda and Nissan were several years behind Daimler's FC knowledge (Kalhammer et al., 1998). Nevertheless, all of the above companies started demonstrating FC models during 1997–1999.[12]

The fuels used in these demonstrations were both hydrogen and methanol, indicating research activities similar to Daimler's.

Furthermore, in press statements, all the above firms indicated that methanol was seriously regarded as a future fuel for FCVs.[13] Several firms, mainly the Japanese (Honda, Toyota, Nissan), indicated that gasoline was not a serious option, given that it would not solve environmental problems.

Kalhammer et al. (1998) note that GM placed emphasis on methanol, whilst keeping the gasoline path open. Nonetheless GM also mentioned the transition advantages of gasoline, namely that gasoline could facilitate a FC breakthrough substantially by not changing both the car technology and the fuel infrastructure. GM actively discussed fuel issues with the oil industry, most importantly with Exxon in 1997. GM's subsidiary Opel, at which the majority of FC activities were carried out until 1998, was more in favour of methanol. This might point to differences of opinion within one (globalized) firm. Opel's inclination towards methanol might have been due to its proximity to technological leader Daimler (Maruo, 1998).

Less pronounced in its fuel strategy was Ford. Ford had developed both hydrogen and methanol-based vehicles at the end of 1999; thereby following the strategy of its FC partner Daimler. But Ford was also discussing the opportunities of gasoline reforming with Mobil.[14] Ford at this time kept its options open. Lastly European companies like Renault, Peugeot, Volkswagen all developed methanol reformers in EU-sponsored programs. Gasoline was not seen as an option for future FCVs (Kalhammer et al., 1998; Maruo, 1998).

The only firms not to develop FCVs on methanol were Chrysler (Kalhammer et al., 1998) and Nissan.[15] By 1998 Chrysler had developed a working prototype of a gasoline reformer, demonstrated in a prototype in 1999. It acknowledged that gasoline reforming was more complex, but the company saw FCVs as a long term technology and thus did not see the gasoline reformer as a delaying factor (Kalhammer et al., 1998). Remarkably, the merger between Daimler and Chrysler led Chrysler to call the gasoline path a developmental dead end.[16] It thereby switched to Daimler's fuel strategy, arguably due to Daimler's favourable position in FC technology, but also due to the more dominant role of Daimler in the merger. Head of the FC program at Chrysler, and fierce protagonist of gasoline over methanol, Mr Borroni-Bird, moved to GM in 1999 (Moltavelli, 2000).

Nissan dismissed both methanol and gasoline. As early as 1997, Nissan stated that it saw more potential in direct hydrogen, and would not develop transition fuels. Nissan's financially problematic situation during this period influenced this dismissal, as it did not have the financial muscle to develop (expensive) reformer technology. Nevertheless Nissan demonstrated a methanol FCV in 2000, indicating a shift within the firm.[17]

3.3.3 Period 3: 2000–2003

In the year 2000 a change can be seen in the fuel preference for FCVs. The change was instigated by GM, which decided to break with the methanol path definitely in 2001. During interviews held with GM in February 2000 there were already indications that fuel preference was shifting towards gasoline. However, the methanol option had not been discarded openly at that time. From May 1997, GM officially partnered with oil companies Exxon and Arco on fuel issues.[18] In May 1999, GM started its partnership with Toyota, generally seen as the closest rival of DaimlerChrysler in FC technology. Meanwhile GM played catch-up by investing heavily in FC technology, a task that Daimler left to Ballard.

From the beginning of 2000, GM showed increased confidence in its own FC technology and according to industry experts was not far behind Ballard's performance.[19] It is in this period of increased confidence that GM announced it would be developing a gasoline reformer with Exxon (September 2000). Not long thereafter in February 2001 GM, Toyota and ExxonMobil agreed that a reformulated gasoline was the preferred fuel for FCVs in the short term, hydrogen in the long term.[20] Arguments included the continuation of the current infrastructure, high total energy efficiency, global availability of gasoline, and consumer acceptance. H. Pearce, GM Vice Chairman, stated in August 2000 that 'the gasoline processor could be the bridge between today's conventional vehicles and tomorrow's hydrogen fuel cell vehicles'.[21]

In the same period GM published a sponsored report in which gasoline-fuelled FCVs showed similar environmental promise to methanol-fuelled FCVs.[22] In the report, GM concludes that methanol would have no advantages over gasoline with respect to efficiency and emission levels in FCVs. Furthermore, GM scrutinized health and safety issues of methanol, and officially stated that the methanol path was not to be considered any more in GM. With regard to commercialization, GM announced in 2001 that it would not strive to be the first on the market with FCVs, but the first car company to sell a million FCVs projected beyond 2008.[23] With GM's and Toyota's fuel strategy so clearly marked, the race for the dominant design for FCVs sharpened.

DaimlerChrysler continued its efforts on the methanol reformer. It received (modest) support from its partner Ford, Mazda (33 per cent owned by Ford), and Mitsubishi (34 per cent owned by DaimlerChrysler). DaimlerChrysler increasingly involved the oil industry for the set-up of a methanol infrastructure, but only found commitment with relatively small players like Methanex, StatOil and BP.[24] DaimlerChrysler seemed to be losing its strong position in FCVs, but nevertheless continued its focus on methanol. Meanwhile

DaimlerChrysler made some agitated remarks concerning competitors' allegations of health and safety problems of methanol. At a conference in Nagoya 2000 a Daimler executive exclaimed: 'Yes, methanol is poisonous if you drink it, but so is gasoline.'[25]

With DaimlerChrysler (and Ford via partnership) on methanol, and GM (and Toyota) on gasoline, the industry was mixed in its fuel preference, and the race to win the dominant design intensified. The gasoline option received a relatively surprising push when the third largest FC consortium in the field (Nissan, Renault and PSA) decided to pursue gasoline FCVs, thereby following GM–Toyota's lead in July 2001.[26] Both Renault and PSA had relatively modest FCV activities, but increased their efforts around 1998–1999, like most of the automotive industry. In July 2000 they decided to join forces, as individual efforts were bound to be marginal given their respective budgets. With this collaboration they formed a counterweight to Daimler–Ford–Ballard and GM–Toyota. With their decision to support the gasoline path the majority of the industry supported gasoline.[27]

As for the rest of the industry, not all companies were that pronounced about their fuel preference. The fourth major player in FC technology was Honda, which had not given any comments on the methanol–gasoline debate, other than that it would continue its direct hydrogen option. All other players are relatively small. Companies like Mitsubishi, Mazda, Hyundai, Daewoo, Fiat, BMW each had their respective FC program, but quite modest. These players either did not have the resources to develop all the necessary components for FCVs, or were partly owned by a parent company; usually the fuel strategy of the parent company is followed (Mazda with Ford, Mitsubishi with Daimler, Opel, Suzuki, Isuzu with GM, etc.). The determining factor seems to be the high cost of developing a gasoline or methanol reformer; therefore its development is limited to those companies with a strong financial arm. In the past Nissan had mentioned that this was a main reason not to choose methanol or gasoline. With its Renault–PSA partnership Nissan seems more confident in pursuing these alternative fuel options.[28]

At a conference in Stuttgart in October 2002, DaimlerChrysler's head of FC activities, Dr. Panik, announced that DaimlerChrysler was continuing its efforts in methanol, although they recognized that gasoline FCVs were becoming an increasingly supported option by the industry.[29] Industry experts commented that DaimlerChrysler was now more or less alone in its methanol preference, and that most companies were either supporting the gasoline path, or remaining with direct hydrogen. Only firms related to or owned by Daimler, mainly Chrysler and Mitsubishi were still active in methanol.

3.3.4   Concluding 1990–2002

Table 6.2 shows the individual firms' fuel use in demonstration vehicles, as an indication of fuel preference. First it shows that all firms active in FC technology have experimented with both hydrogen and methanol. This reflects the widespread activities in methanol, and indicates the industry consensus on this fuel. Second, it shows the hydrogen models are still in the majority; this reflects the fact that although alternative fuels are explored, hydrogen-based FCVs are technically still seen as the best solution. Third, there are still only two gasoline FCV demonstration models, due to the complexity of the gasoline reformer in comparison to the methanol reformer. In this case press statements are essential in assessing fuel preference. The variable 'number of demonstration vehicles' thus has some limitations; company statements serve as a necessary addition. In the future patent research could further enhance the assessment of fuel preference.

Figure 6.3 shows the accumulated number of demonstration vehicles using either hydrogen, methanol or gasoline over time, quarterly. It shows how hydrogen FCV dominated until the beginning of 1997. Methanol FCVs increased from then until the end of 2000, when this number stabilized. From that point on hydrogen FCV surged.

Table 6.2   Fuel preference in demonstration vehicles per firm 1993–January 2002

Firm	Hydrogen models	Methanol models	Gasoline models
DaimlerBenz/Chrysler	6	2	–
General Motors	3	3	1
Toyota	5	2	–
Ford	3	1	–
Honda	3	2	–
Chrysler (until 1999)	–	1	1
Nissan	1	1	–
Mazda	2	1	–
Renault	2	–	–
PSA	2	–	–
BMW	2	–	–
Hyundai	1	1	–
Mitsubishi	1	1	–
Volkswagen	2	–	–
Fiat	1	–	–

fuel preference in demonstration vehicles

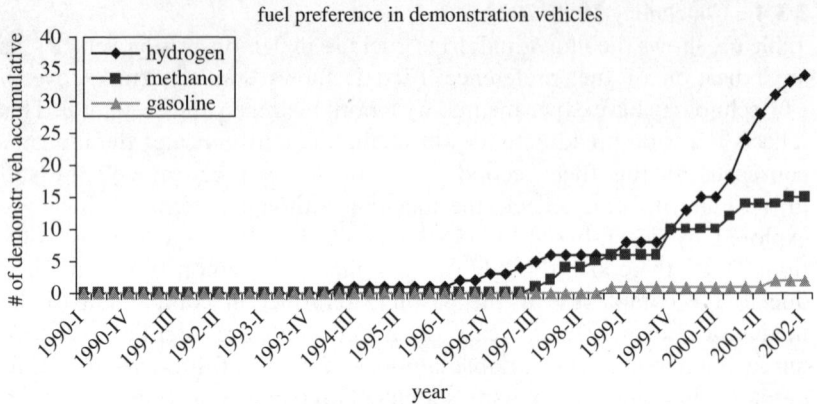

Figure 6.3 Industry fuel preference in demonstration vehicles 1990–2002

Figure 6.4 schematizes the fuel preference of the different car companies over the decade. Methanol dominated in 1997–1999. A shift to gasoline can be discerned from then on. In the patterned area underneath in Figure 6.4 the companies are mentioned which have not made specific announcements on preference; however industry experts expect most of these companies to have discarded methanol (Volkswagen, Honda, Ford) and to be actively evaluating gasoline (Ford, Mazda).

4.   ANALYSIS

Based on the above description of the industry reaction to fuel preference the process of technological decisionmaking and determinants of technological change will be discussed.

4.1   Technology Choices at Industry Level

How do technology choices at industry level come about? The case shows how within the three periods one fuel dominates as the preferred fuel within the automotive industry to fuel FCVs. Given that the hydrogen preference in the early 1990s is due to the experimental nature of FC research, the intentional choices for methanol and gasoline as the preferred fuel in the commercial FCV are most interesting to study. The following mechanisms can be discerned for methanol as well as gasoline.

First, a *'credible actor'* in the automotive industry proposes an alternative to the current dominant technological solution; the credible actor plays

	Hydrogen-fuelled FCV	Methanol-fuelled FCV	Gasoline-fuelled FCV
'90–'96	Daimler Benz GM Toyota Ford Mazda Chrysler		
'97–'99	Nissan Mazda BMW	Daimler Benz GM Ford Toyota Honda Renault PSA Hyundai Volkswagen	Chrysler (–98)
'00–'02	BMW	DaimlerBenz/Chrysler Ford Mitsubishi Mazda Hyundai Honda Volkswagen Fiat	Toyota GM PSA Renault Nissan

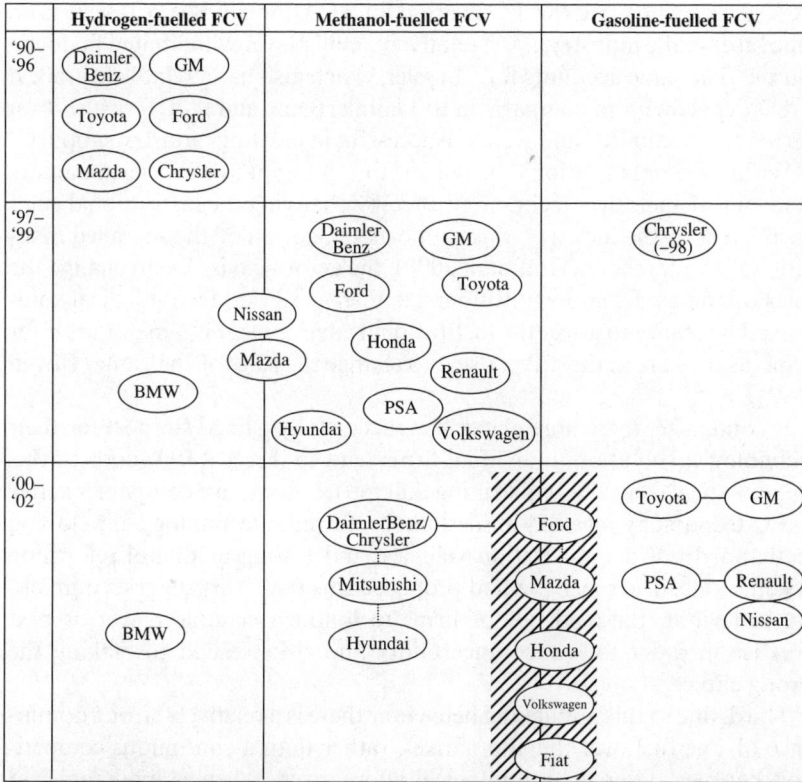

Figure 6.4   Shifting FCV strategy of the main car manufacturers between 1990 and 2002

the role of opinion leader within the industry by deviating from the current 'routine' or 'technological solution'. In the case of methanol the opinion leader was DaimlerBenz shifting from hydrogen; with gasoline General Motors took the lead by deviating from the widely supported methanol.

DaimlerBenz owed its credibility to its leading position in FC technology, its partnership with FC leader Ballard, and its strong name in the industry as an automotive innovator. General Motors owed its credibility to its market leadership and financial arm, combined with its partnership with Toyota (globally third automotive manufacturer) and Exxon (market leader in the oil industry). Credibility is an important determinant of whether or not the industry will support the suggested new routine. For several years BMW has unsuccessfully promoted hydrogen-based ICEs, and FC technology used a battery replacement. Also Chrysler's attempts to get gasoline

FCV accepted in the 1996–1998 period failed. Although BMW is seen as an innovator in the industry, it is a relatively small player with limited financial muscle. The same accounts for Chrysler, which also has a relatively modest R&D department in comparison to DaimlerBenz, and its US rivals. Both lacked the credibility, and were unsuccessful in creating industry support.

Within the organizational field firms thus attempt to win relevant actors in favour of their own technology, thereby changing the institutional rules for the rest of the industry; this supports the notion of the so-called institutional entrepreneur (Hoffman, 2000), the actor who is able to change the rules of the game, and institutionalize their preferred technological solutions. The ability to affect the institutional environment is a major asset for firms, as they are in the driver's seat to change the rules of the game (Oliver, 1992).

Second, after a credible player has successfully gained support for their technological solution, individual firms tend to direct R&D funds to this technological solution. It is striking that most automotive companies active in FC technology invested in methanol reforming technology, developing methanol FCV demonstration vehicles, and testing methanol reforming systems, once DaimlerBenz had proposed this fuel. This suggests mimicking behaviour: the tendency of firms to follow a credible player of best practice in order to reduce uncertainty and risks related to making the wrong choice.

Third, due to this mimicking behaviour there is a relatively strong dominance of one fuel over the alternatives, rather than a continuous competition between different technological alternatives. Whereas methanol had the support of most of the automotive industry around 1999, only DaimlerChrysler was currently continuing this effort at the time of writing. The majority of firms in the industry has shifted to gasoline. Mimicking of credible actors is not the only mechanism by which individual firms choose to conform to a certain technological solution. The nature of the industry, changed through mergers and acquisitions, has an important influence as well.

More or less *dependent automotive companies* (subsidiaries or partly owned companies) are more likely to follow the technical choice of the parent company. In order to strengthen the alliance with respect to its technical choice parent companies convince/coerce subsidiaries and partly owned companies to their preferred position. The best example of this phenomenon comes from Chrysler's shift to methanol when it merged with Daimler, despite Chrysler's efforts to develop gasoline reformers. Alliance pressure was thus dominant over internal beliefs that gasoline made more sense.

Independent automotive companies in an FC alliance (technology-related alliance) are likely to support the technological decisions of the dominant alliance player. Examples include Ford's support to Daimler's methanol path, and Toyota's support to GM's gasoline shift. This is however not a form of coercion or mimicking. This seems largely motivated by strategic motives of strengthening the alliance and its technical choices.

4.2 Opinion Leaders and Technological Deviations

If opinion leaders are indeed so important in technological choices, as is suggested in the above, then which factors determine their specific choices? More specifically: do these factors originate from the institutional environment, or more from firm-internal motives? Are these choices technology-specific, or are political aspects more dominant?

4.2.1 DaimlerBenz/Chrysler and methanol
In 1996 DaimlerBenz could broadly choose between methanol and gasoline (ethanol and natural gas have never been actively discussed and developed by the industry), and preferred the first. At this point in time there were no rules or standards to conform to; FC technology was a new technology, the dominant design was not yet set (no normative rules), and to the extent that design decisions were already set, they were set by DaimlerBenz itself. The technology was unfolding, in its possibilities as well as its bottlenecks. Within this context, DaimlerBenz had the freedom to choose, and with its competitive advantage over its competitors it had the opportunity to shape the dominant design in line with its interests.

As said *functional aspects* played an important role in the decision to move to methanol, and not gasoline. Methanol formed 'the best option' for the job (easy reforming, relatively efficient, and a sustainable solution): DaimlerBenz considered these arguments strong enough to convince the oil industry to develop a methanol infrastructure. Related is *the environmental image* of FC technology. The FC program within DaimlerBenz had been set up with the specific environmental problems associated with the ICE. Shifting to gasoline would undermine the very reason for FC activities.

Apart from functional aspects, *strategic motives* played an important role. Choosing gasoline would have set the deadline back for the introduction by a number of years. In that period Daimler might lose its competitive position; the window of opportunity was there for Daimler to use.

Lastly, the methanol choice also reflects *differences in institutional context* between Europe and the USA. The issues of the 'greenhouse effect' and 'renewable energy' are higher on the agenda in Europe (and Japan) than in the USA. Both in Europe and Japan methanol was preferred due to the opportunity it offered to reduce greenhouse gases and form a sustainable energy source (when produced from biomass). In contrast, all US companies had a past in gasoline reforming: GM with Exxon, Ford with Mobil, and Chrysler before it merged with Daimler. The combination of strong influence of the oil industry, as well as the lower priority of greenhouse issues seems to explain differences in fuel choices between GM and DaimlerBenz. DaimlerBenz gave higher priority to functional characteristics and the intrinsic advantages of methanol over gasoline, rather than to the explicit rejection of methanol by the automotive industry.

4.2.2   General Motors and gasoline

When fuel preference became an issue around 1997–1998 GM followed Daimler's preference for methanol, due to Daimler's lead over GM in FC technology. Furthermore, GM's FC research was carried out in Mainz, Germany at its Opel subsidiary. In the past Opel made more positive statements about methanol than GM itself. GM's methanol preference might thus also reflect Opel's preference.

GM gave *functional arguments* for gasoline over methanol given the problems associated with developing a methanol infrastructure, as well as technical advantages. Gasoline is widely available; it has a relatively high energy density in comparison to methanol (resulting in more range); and there are no health and safety problems associated with gasoline (methanol is transparent). However, Daimler also used functional arguments that methanol made more sense: apparently the priorities in these arguments differed between Daimler and GM.

The *oil industry* seemed to have played a dominant role in GM's shift. Exxon mandated gasoline, and dismissed methanol. Furthermore, methanol would require alterations in the current infrastructure, as the established tanks (at tank stations) would contaminate the methanol. GM acknowledged that without the oil industry infrastructure would be lacking for methanol FCVs, and thus discarded this option. Less emphasized is that apart from methanol, the projected gasoline for FCVs would also require new storage tanks to counter contamination of the clean fuel: the costs of developing an infrastructure for clean gasoline or methanol are similar.

Another point suggesting why GM was more inclined to shift than Daimler relates to *strategic factors*. Daimler had a clear lead in methanol reforming technology. In a period in which methanol became more scrutinized (2000) choosing gasoline would undermine Daimler's competitive position in methanol further. The support of Toyota and Exxon was crucial to give this shift sufficient push and credibility, as well as its own strong position in FC technology which it had acquired in recent years.

Another *strategic factor* is that choosing gasoline would permit automotive firms more time for the development of FCVs; it would delay the commercialization of FCVs by several years. Although GM has announced that it strives to be the first car maker to sell one million FCVs, it has historically been sceptical about the potential and expectation of FC technology, being complex and expensive. The FCV program plays an important legitimizing role for automotive firms towards regulators and Californian regulators in particular, demonstrating the good will to address environmental issues. Postponing fits with a decade-long strategy of litigation and confrontation with regulators with regard to new emission and energy efficiency standards.

Lastly, as said earlier the *institutional context* favours gasoline over methanol in the USA, due to different priorities assigned to renewable resources and greenhouse gas effect.

Concluding, one can say that the case of fuel preference and Daimler's and GM's respective choices of methanol and gasoline are determined by a combination of institutional differences, technological (functional) characteristics, and firm-specific interests.

The case also shows that neither methanol nor gasoline can be called 'the best option': both have their specific merits and bottlenecks; firm-internal beliefs and convictions seem important to prioritize among the specific qualities of each fuel. Daimler pursued an environmental technology, based on its environmental image and sense of urgency felt with the greenhouse gas issue; GM stressed transition advantages of gasoline, thereby following its own conviction that gasoline was not the problem.

Lastly strategic motives have played a role in two ways: firstly, choosing a certain technology undermines the position of competitors, and secondly the timeline of innovation was influenced.

5.   CONCLUSIONS

In this chapter a specific technological choice process is described with regard to the preferred fuel for the future fuel cell vehicle (FCV). The case provides insights into the process by which technological decisions are made within an industry concerning new technologies, and how certain technological trajectories are terminated. In this case the choice of methanol around 1997, and the shift from methanol to gasoline around 2000 provide relevant insights into this process.

First, the case suggests that periods in which technological options dominate succeed each other. Methanol was the preferred fuel of most of the automotive companies in the late nineties, until a shift to gasoline took place, which is now the dominant option for fuelling FCVs.

Second, it is useful to make a distinction between opinion leaders and the 'followers' of the industry. The case shows how two or three opinion leaders in the automotive industry have the credibility to shape the discussion on the fuel, and deviate from the established route. Credibility is based on a strong resource position, a strong network or a historically built name as an innovator. Credible players or coalitions are able to gain support for their option from industry players, leading to mimicking behaviour, which in turn leads to an institutionalization of the proposed solution direction. This takes the form of rules and norms (normative pressure), but can in time lead to regulatory standards. Credibility is an important asset enabling firms to change the rules of the game, or change the institutional context. The case thus provides some evidence of the institutional entrepreneur.

Third, as for the 'followers' in the industry, mimicking behaviour can be discerned on an organizational level with the majority of firms: in times of uncertainty over future technological directions, best practice is followed in order to reduce uncertainty, risk and search costs. Apart from mimicking behaviour, coercive pressure can be observed when parent companies mandate their preferred fuel on their subsidiaries or their weaker alliance partners.

Fourth, with respect to how technological decisions are made, opinion leaders are influenced by a number of factors with no clear dominance. A combination of institutional differences, strategic motives, and internal motives (defining technological priorities) are important determinants in the choice process.

NOTES

1. DaimlerBenz and Chrysler merged in 1998 to form DaimlerChrysler.
2. Platform strategy refers to the specific focus on developing a limited set of platforms, on which a multitude of models can be built. Since Volkswagen introduced this strategy in the 1980s it has become widespread in the industry.
3. www.fuelcells.org/fcnews.htm (see January 1997).
4. www.fuelcells.org/fcnews.htm (see April 1997).
5. FC investments are estimated based on press releases of car companies during 1995–2002, combined with company reviews selected from the Business and Industry database, investment and data quoted for 2002, www.gm.com, www.toyota.com, www.ford.com, www.daimlerchrysler.com.
6. Data for 2002.
7. Based on interviews with DaimlerChrysler.
8. Based on interviews with DaimlerChrysler; press release DaimlerBenz, May 1997.
9. Based on interviews with DaimlerBenz.
10. www.fuelcells.org/fcnews.htm (see February 2000).
11. Daimler did not have to invest in BEV technology as it was not affected by the ZEV requirements being a small scale manufacturer in California, and it could allocate all resources to the fast moving FC technology.
12. See www.fuelcells.org, monthly newsletter in this period.
13. See press releases by Toyota (December 1997), General Motors (April 1998), Nissan (September 1998), Honda (November 1999), and Ford (December 1999).

14. www.ford.com, downloaded April 2002.
15. Based on interviews.
16. www.fuelcells.org/fcnews.htm (see January 1999).
17. Nissan interviews.
18. www.fuelcells.org/fcnews (see May 1997).
19. Interview JARA, Japan Automobile Research Center, November 2000.
20. Exxon and Mobil merged in 2000.
21. www.evworld.com, August 2000.
22. www.fuelcells.org/fcnews (see April 2001).
23. GM press release, March 2001.
24. The partnership with BP did not involve methanol infrastructure development, but focused on hydrogen.
25. DaimlerChrysler, November 2000.
26. www.fuelcells.org/fcnews (see July 2001).
27. GM, Toyota, Renault–Nissan–PSA have relatively large FC programs in comparison to the rest of the industry. Only DaimlerChrysler, Ford and Honda have similar budgets: The gasoline supporting alliances thus represent the majority of the industry in terms of FC budgets.
28. Renault and Nissan took over when Renault bought 37 per cent of the stock of Nissan in 1998.
29. F-cell conference, Stuttgart, 27–28 October 2002.

Table A6.1   Firms interviewed

Firm/country	Number of interviews	Period of interviews	Department /position
General Motors, USA	8	February 2000	Global Alternative Propulsion Centre/ director
			Chemical and Environmental Science Laboratory/Director, Senior Researcher, Principal Research Scientist, Manager Fuel Chemistry and Systems
DaimlerChrysler, Germany	3	October 2000	FC division / Assistant Senior Project Manager, Communications
			R&D department / Senior Researcher
BMW, Germany	3	April 2001	Energy and Drive Train Research / Head Electric Systems
			Energy and Environment / Researcher Energy and Environment / Public Relations

Table A6.1 (continued)

Firm/country	Number of interviews	Period of interviews	Department/position
Nissan Motor Co., Japan	2	November 2001	Nissan Research Center / Research Director
			Environmental and Safety Engineering Department / Senior Managerial Specialist
Honda Motor Co., Japan	3	November 2001	Corporate Planning Division / Chief Engineer, General Manager
			Corporate Communications Department / Manager
Renault, France	2	February 2002	R&D department / Group Head, Researcher

REFERENCES

Brint, S. and J. Karabel (1991), 'Institutional origins and transformations: the case of American community colleges', in W. W. Powell and P. J. Dimaggio (eds), *New Institutionalism in Organizational Analysis*, Chicago: University of Chicago Press, pp. 337–60.

DiMaggio, P. J. and W. W. Powell (1983), 'The iron cage revisited: institutional isomorphism and collective rationality in organizational fields', *American Sociological Review*, **48** (April), 147–60.

Dosi, G. (1982), 'Technological paradigms and technological trajectories', *Research Policy*, **11** (3), 147–62.

Giddens, A. (1984), *The Constitution of Society: Outline of the Theory of Structure*, Berkeley, CA: University of California Press.

Greenwood, R. and C. R. Hinings (1996), 'Understanding radical organizational change: bringing together the old and the new institutionalism', *Academy of Management Review*, **21** (4), 1022–54.

Hart, D. and A. Bauen (1998), *Further Assessment of the Environmental Characteristics of Fuel Cells and Competing Technologies*, London: Energy Technology Support Unit, Department of Trade and Industry.

Haveman, H. A. (1993), 'Follow the leader: mimetic isomorphism and entry into new markets', *Administrative Science Quarterly*, **38**, 593–627.

Hoffman, A. J. (2000), *From Heresy to Dogma: An Institutional History of Corporate Environmentalism*, Boston, MA: Stanford Business Books.

Höhlein, B. (1998), 'Vergleichende Analyse von Pkw-Antrieben der Zukunft mit verbrennungsmotoren oder Brennstoffzellen-Systemen', *VDI Berichte*, Munich.

Kalhammer, F. R., P. R. Prokopius, V. P. Roan and G. E. Voecks (1998), *Status and Prospects of Fuel Cells as Automobile Engines: a Report of the Fuel Cell Technical Advisory Panel*, Sacramento, CA: State of California Air Resources Board.

Levy, D. L. and S. Rothenberg (2002), 'Heterogeneity and change in environmental strategy: technological and political responses to climate change in the automotive industry', in A. J. Hoffman and M. J. Ventresca (eds), *Organizations, Policy and the Natural Environment: Institutional and Strategic Perspectives*, Stanford, CA: Stanford University Press.

Maruo, K. (1998), *Strategic Alliances for the Development of Fuel Cell Vehicles*, Goteborg: Göteborgs Universitet.

Moltavelli, J. (2000), *Forward Drive: the Race to Build 'Clean' Cars for the Future*, San Francisco, CA: Sierra Club Books.

Nelson, R. R. (1991), 'Why do firms differ, and how does it matter?', *Strategic Management Journal*, **12** (Winter), 61–74.

Nelson, R. R. and S. G. Winter (1982), *An Evolutionary Theory of Economic Change*, Cambridge, MA: Harvard University Press.

Oliver, C. (1992), 'The antecedents of de-institutionalisation', *Organization Studies*, **13** (4), 563–88.

Oliver, C. (1997), 'Sustainable competitive advantage: combining institutional and resource-based views', *Strategic Management Journal*, **18** (9), 697–713.

Pfeffer, J. and G. Salancik (1978), *The External Control of Organizations*, New York: Harper & Row.

Scott, W. R. (1995), *Institutions and Organizations*, London: Sage.

Van den Hoed, R. and M. Bovee (2001), 'Responding to stringent environmental regulation: the case of the automotive industry', Eco-Management and Auditing Conference proceedings, Nijmegen, 21–22 June.

Van den Hoed, R. and P. J. Vergragt (2001), 'Explaining environmental technology strategies in the automotive industry: the case of fuel cell technology', Greening of Industry proceedings, Bangkok, 21–25 January.

7. Distant networking? The out-cluster strategies of new biotechnology firms

Margarida Fontes

1. INTRODUCTION

The biotechnology industry is characterized by a network structure of interorganizational relationships that acts as a coordination device between a variety of actors – new biotechnology firms, large established firms, universities and other nonfirms – with diverse competencies and assets (Barbanti et al., 1999; Powell et al., 1996). Given this specific form of industrial organization, the spatial concentration of innovative activities was found to favour biotechnology development, with location in 'biotechnology clusters' emerging as a factor of firms' competitiveness (Cooke, 2001). However, there is evidence too that biotechnology firms are also more likely to establish connections outside the regional environment, given the global nature of their markets and the diversified and fast changing nature of the science base needed to innovate (McKelvey et al., 2003). In this context, the success of new biotechnology firms, which perform an intermediate function between science and the market, depends on their ability to put together a coherent set of relationships, both close by and distant, that enable access to new scientific knowledge and to the establishment of effective channels to technology or product markets.

Against this background it is possible to ask the question: how do firms operating outside biotechnology clusters manage to survive and develop? The objective of this chapter is to address this question by identifying and discussing the main features of an 'out-cluster' strategy, based on case studies of Portuguese new biotechnology firms.

This chapter is arranged as follows. Section 2 examines the network structure of the biotechnology industry, focusing on the role of the new biotechnology firms and the importance of local versus distant relationships. After identifying the features that allow distant relationships (and thus peripheral locations) to be viable, six propositions regarding the relative importance of and establishment of regional, national and international relationships are derived. Section 3 outlines the methodology and describes the general characteristics of firms interviewed for this study. Section 4 examines the structure and composition of firm relationships and analyses the underlying firm strategies and motives. The section focuses, particularly, on the strategies of biotechnology firms operating outside established clusters. Section 5 characterizes 'distant networking strategies' and discusses two patterns in the establishment of foreign technological relationships. Section 6 concludes.

2.   NETWORKS, PROXIMITY AND DISTANCE IN BIOTECHNOLOGY

2.1   The Network Structure of the Biotechnology Industry and the Position of New Firms

According to some authors the abstract and codified nature of scientific knowledge introduces a certain linearity in the innovation process in biotechnology, permitting its separation into stages and thus favouring a division of labour, where universities and other public sector organizations would be concerned with the production of new scientific knowledge, new biotechnology firms (NBFs) with the transformation of this scientific knowledge into technological and commercial applications and large firms with the production and marketing activities (Arora and Gambardella, 1994). However, this 'division of labour' is not static and has registered some changes through time, driven by the interactions between actors and by alterations in the properties of the underlying technologies, which led to some modifications in the patterns of specialization (Barbanti et al., 1999; Orsenigo et al., 2001; Queré and Saviotti, 2002). NBFs maintain a critical role, by conducting a transformation process that enables the mobilization and productive use of knowledge generated in research organizations (Fontes, 2001). Still, there have been some changes in their actual functions, from acting as translators between research organizations and large firms still building a knowledge base in the new field to acting as explorers of new and/or alternative paths, enabling the large firms to explore a wider range of technological approaches (Pyka and Saviotti, 2000).

The role played by NBFs is based on their ability to gain access to and identify application opportunities for new knowledge originating from research. But, NBFs specific competences and organizational structure also present some disadvantages both in terms of breadth of knowledge – they can be too specialized to operate independently from large firms – and regarding the downstream stages of the innovation process: compliance with regulatory requirements, large scale production and commercialization, where competences and assets often lie with large established firms (Arora et al., 2001; McKelvey and Orsenigo, 2001).

Therefore, the NBFs' capacity to explore their particular type of advantages depends on their ability to access both new scientific knowledge and markets channels. We will subsequently address the conditions in which firms created outside major biotechnology agglomerations can compensate for the limitations of their local environments in these areas.

2.2 Clustering and Reaching Out in Biotechnology

Research on the behaviour of high technology firms located in regions where knowledge accumulation is lower, although scarce, shows that successful firms will reach out for knowledge and resources they cannot find locally and therefore will tend to rely more frequently on distant relationships (Cooke, 2001; Felsenstein, 2001; Rees, 2001; Saxenian and Hsu, 2001). Echeverri-Carroll and Brennan (1999) conclude that the importance of proximity is relative, depending on the local accumulation of knowledge and that when such accumulation is smaller, firms will look for knowledge elsewhere, where it is available.

Biotechnology is likely to be one field where this type of behaviour may emerge because of some features of biotechnology, related to the nature of knowledge production and the characteristics of markets, may facilitate the development of distant strategies. We will now discuss this issue in detail.

2.2.1 The relative importance of local vs. distant relationships in biotechnology

Evidence from the USA and Europe shows that biotechnology firms appear to benefit from being located in strong regional clusters and simultaneously from being positioned in transregional networks that enable them to interact with a greater variety of organizations and to access a wider range of competencies and resources (Allansdottir et al., 2002; Cooke, 2001; Owen Smith et al., 2002). In fact, biotechnology shows a strong tendency towards clustering, which is associated with: the quality, variety and level of integration of the science base; the industry's absorptive capacity; and the presence of supporting institutions, namely financial and labour markets. But, there is also a parallel tendency of existing clusters to open up and establish a variety of external connections, associated with the need to access leading edge research (Allansdottir et al., 2002).

Additionally, it has been shown that the relative importance of the cluster is not the same for all activities and partners. For instance, co-location appears to be relatively more frequent in relationships with research organizations (ROs) than with other firms (McKelvey et al., 2003), suggesting that proximity is particularly relevant for knowledge-intensive relationships.

In fact there is evidence that firm-to-firm linkages (which are potentially more market-oriented) are more frequently established outside the regional environment (McKelvey et al., 2003; Swann and Prevezer, 1996), which may be explained by the global nature of markets for biotechnology (Cooke, 2002). However, the NBFs' capacity to enter global markets depends on the nature of their business: those developing unique technologies and products are less dependent on the local environment than service firms or suppliers of intermediate products (Mangematin et al., 2002; Zeller, 2001). Also, the capacity to engage in these more ambitious and risky market strategies is greatly influenced by the availability of financial resources (Mangematin et al., 2002). Because venture capitalists tend to invest more in firms located nearby, it may be more difficult to secure distant investors (Stuart and Sorenson, 2003). Moreover, NBFs may also need market-oriented partners for the co-development and downstream innovation stages. This type of relationship can be more complex to establish at a distance by firms without a previous track record and can be harmed by negative country-of-origin effects (Roth and Romeo, 1992).

Concerning the capacity to engage in international activities at an early stage, the literature on 'born global' firms provides some indication of the factors that can assist it: the entrepreneurs' previous international experience and the quality of their personal network; ownership of specialized and difficult to imitate resources; and the capacity to engage in alliances and cooperative networks (Coviello and Munro, 1995; Crick and Jones, 2000).

Finally, local relationships tend to be more important in the early stages, becoming less relevant during the firms' life cycle (Lemarié et al., 2001). Biotechnology entrepreneurs, who often originate from ROs, tend to locate their firms in the neighbourhood of their source organization, for reasons such as advantages in the access to knowledge, risk reduction strategies, personal mobility and informal networking (Breschi and Lissoni, 2001; Lemarié et al., 2001; Stuart and Sorenson, 2003). However, the relevance of co-locating is greater for junior scientists, who may be more dependent on the 'parent' RO, than for senior scientists with an international reputation (Mangematin et al., 2002). It will also depend on the type of assets the firm wants to obtain (Audretsch and Stephan, 1996): credibility or occasional assistance vs. vital knowledge, whose access may require co-location to where key scientists are (Zucker et al., 1998).

Since knowledge appears to be a particularly important asset for NBFs and one that most literature stresses as being difficult to access at a distance, it is relevant to understand whether and in what conditions firms located outside the main biotechnology concentrations can obtain this critical input.

2.2.2   Access to knowledge at distance in biotechnology

One of the advantages of locating in a region where knowledge accumulation is higher concerns the transmission of knowledge, particularly of tacit knowledge (Feldman, 1999). Biotechnology relies extensively on scientific knowledge which is, in principle, more abstract and codified (Arora and Gambardella, 1994) and thus more easily transmitted at a distance, especially when access to distant information has become easier and more affordable (Amin and Cohendet, 2003).

However, tacit knowledge still plays a very important role, especially in the early stages, given the 'natural excludability' of new scientific discoveries (Zucker et al., 1998). This means that only those who were involved in the development of the technology, or have direct access to the research team who did it, will possess the knowhow necessary to replicate this knowledge. People who had such a common experience may have developed shared meanings, a shared language and communication codes – i.e. epistemic proximity (Steinmueller, 2000). This creates conditions which allow for the knowledge produced to be at least partly articulated and transmitted at a distance between members of the same 'epistemic community' (Breschi and Lissoni, 2001).

Co-location is necessary for the co-development and creation of epistemic proximity. But, while exploration activities (co-production of new knowledge) will require a more permanent co-location, exploitation activities (e.g. absorption and recontextualization of the knowledge produced) only require temporary co-location (Gallaud and Torre, 2001). Also, co-location of people is not necessarily synonymous with the co-location of firms, although small firms and start-ups may find the latter more favourable (Gallaud and Torre, 2001). Other forms of the co-location, like postgraduate studies or periodical stays in a research lab may create the conditions for shared experiences, even if the actor originates from a distant region (Amin and Cohendet, 2003). Additionally, 'virtual' communication can be instrumental in facilitating the working of scientific communities and to assist members in less central positions to establish and maintain an effective collaboration (Walsh and Bayma, 1996).

There are nevertheless some nonreproducible advantages of locating in environments where research is world class. Embeddedness in local social networks facilitates access to information on a 'who knows what' and 'who does what' basis, which can trigger early contacts (Breschi and Lissoni, 2001), which are particularly important when the new knowledge being searched is not publicly available (Arundel and Geuna, 2001). Finally, some authors argue that the proximity achieved by belonging at a distance to a 'community' has its limitations, because virtual communications do not 'offer the same scope for reciprocity, serendipity and trust that is afforded by sustained face to face contact' (Morgan, 2001, p.15; Roberts, 2000).

2.2.3 The nature of distant search

Given the importance of scientific advances in biotechnology and the international nature of knowledge production, the search for knowledge outside the regional environment may be required, even for firms located in a biotechnology cluster. However, it can be argued that, while the latter will more frequently look for nonredundant knowledge that enables them to renovate or reconfigure their knowledge base and to avoid the risk of excessive in-breeding (Bathelt et al., 2002; Rosenkopf and Almeida, 2003), firms located outside the main knowledge concentrations will first of all look for knowledge that enables them to develop and exploit the existing knowledge base and only later will eventually start looking for other types of inputs. For this reason, at least in the early stages, 'out-cluster' firms are likely to search for knowledge that is not too far from their current knowledge base. Given the path-dependent nature of innovation, firms will find it easier to rely on their existing knowledge base to conduct new searches and will be more able to understand and absorb knowledge that is closer to it (Cohen and Levinthal, 1990). Their search will therefore rely on technological proximity at a geographical distance. Later, firms may reach a point when they also need to look for substantially new knowledge, meaning technological and geographical distance (Rosenkopf and Almeida, 2003). Here, their experience in establishing distant relationships can be an asset.

Finally the search for relationships at a distance presents some particular features (Bathelt et al., 2002; Lorenz, 1999). It is more purposeful and focused, because it does not come occasionally or without costs, but rather is the result of a conscious effort to identify and gain access to a particular type of partner; usually trust does not exist at the outset and has to be built; it can be a slow process, with firms tending to apply staged procedures, where levels of risk and commitment from the partners increase through time. Because these relationships take more time and effort to establish and maintain, there are tighter limits upon the number of linkages firms are able to manage.

The above discussion enables us to confirm that some features of biotechnology may indeed favour the development of strategies that do not rely strongly on the advantages of proximity. The global nature of biotechnology markets may enable (precocious) market internationalization, although the chances of success depend very much on the nature of the technologies, the resources available and the background of the entrepreneurs. On the other hand, while access to technological knowledge is critical and the transmission of knowledge (particularly tacit or 'excludable' knowledge, or knowledge that is more remote from the firms' knowledge base) can be complex, these difficulties can be circumvented or lessened, in some conditions. For instance, new knowledge can be more easily transmitted at a distance between actors that were involved in processes of co-production, which provides for epistemic proximity. While the latter may require co-location, it does not necessarily mean the co-location of firms, entrepreneurs and new firms may profit from alternative forms of co-location for their purposes. Moreover, not all required knowledge will necessarily be frontier knowledge and thus the search for it may be possible through various sources of information, providing that it is not too far from the firm's existing knowledge base.

In any case, the search for distant relationships and the process of their establishment have a particular pace and requirements, which have to be taken into account by firms relying extensively on this type of process. Indeed, it is important to have in mind that location in a biotechnology 'cluster' can confer advantages in this industry and thus firms located 'out-cluster' are likely to face and need to overcome specific difficulties.

2.3 Distant Networking Strategies – Propositions

The above discussion provided an overview of features of biotechnology and biotechnology firms that may enable these firms to operate at a distance from the main centres of competence accumulation in biotechnology. When combined with the existing knowledge concerning the development of biotechnology in a 'peripheral' region (Fontes, 2001; Fontes and Pádua, 2002), they lead us to argue for the viability of a strategic behaviour characterized by the search for external resources and competencies where they are available (nearby or distant), which inevitably means extensive sourcing outside the regional environment and a precocious internationalization along several dimensions. The need to comply with different conditions and to adjust to different structures of relationships leads these firms to display behaviours that are distinct from those of similar firms in other types of environments.

On the basis of this reasoning we put forward the notion of 'distant networking strategies', which are likely to typify the behaviour of NBFs in this type of environment, and we advance a number of propositions, as a first contribution to characterizing them.

The first set of propositions regards the relative importance of regional/national and transnational relationships:

Proposition 1 – Firm formation decisions are associated with the presence of sources of scientific knowledge, particularly local/national research organizations (ROs), but . . .

Proposition 2 – Firms/entrepreneurs already have at the outset, or build at an early stage, a greater or lesser range of external connections to access complementary knowledge, which will strongly influence the knowledge base of the new firm.

Proposition 3 – Firms will be required to search externally for markets and market-related linkages, given the limitations of local demand and therefore will be engaged in processes of precocious market internationalization.

Proposition 4 – International relationships will become progressively more important along the firms' life cycle, as they move towards commercialization stages and/or need to broaden or renovate their knowledge base.

Two further propositions address the way firms set about to establish distant relationships:

Proposition 5 – Access and establishment of scientific and technological relationships will be more frequently mediated, both by local scientific networks – that act as gateways to international ones – and by personal networks, given the frequent scientific background of firm entrepreneurs and employees.

Proposition 6 – There will remain an important element of 'unsupported' search for relationships, particularly market-oriented relationships, but also scientific and technological ones in fields where local competencies are absent. In these conditions the identification of potential partners, credibilization and acceptance are likely to be more complex and to require specific mechanisms to be achieved.

These propositions will guide our empirical research on the actual behaviour of Portuguese NBFs.

3.   EMPIRICAL RESEARCH: METHODOLOGY

In order to obtain an in-depth understanding of the strategies adopted by 'out-cluster' firms, empirical research has been conducted on a group of Portuguese biotechnology firms.[1] The Portuguese environment is not particularly favourable to entrepreneurial initiatives in this field (Fontes, 2001) and as a result, there are very few new biotechnology firms in Portugal, most of them very recent: of the 29 companies that are still active, half were created after 2000. But despite these numbers, there are a few older firms that have had some success and have reached a stage where it is possible to look back at their development process. Six of these firms were chosen for this analysis: the four older surviving firms (created between 1990 and 1996), to which two younger firms were added (created in 1998 and 1999), to have the counterpoint of firms going through early development stages in a more recent period.

The empirical analysis combined previous accumulated knowledge about these six companies (all were the object, through time, of periodical follow-ups, enabling a quasi-longitudinal view of their evolution), with purposeful data collection on their linkages and in-depth interviews conducted with a view to obtaining detailed information about firms'/entrepreneurs' rela-tionships (formal or informal): establishment, management, motivations, difficulties and underlying strategies.

The research involved: mapping the network of NBFs' technology and market-oriented relationships and assessing the importance attributed to and the roles performed by the different elements; evaluating the relative importance of regional/national vs. transnational linkages; understanding the strategies (articulated or not) underlying the choices made by firms regarding the structure of their relationships and the motives behind the definition of these strategies; analysing the process of establishment and management of distant relationships and discussing the implications of operating at a distance from relevant knowledge centres and main markets, as perceived by the firms.

Given the small number of biotechnology firms in Portugal it will be rela-tively easy for someone in the milieu to identify the firms that were the object of the case studies. For this reason it was decided to avoid providing specific data about each firm, for the various themes addressed (although the data were systematized and analysed in detail) and thus only general conclusions and global patterns are presented.

Table 7.1 gives an overview of the firms studied. These firms are at different stages of business development, which is obviously related to age and to the type of technology being developed. But it is also influenced by the level of resources available, which may enable concentration on the main business or force the dispersion by other activities, delaying development (a case limit is firm C, that started up in 1994 but for financial reasons only focused decisively on biotechnology activities in 1999). They operate in a variety of application fields: pharmaceuticals, environment, marine prod-ucts, food processing, with some targeting more than one field.

All firms were direct or indirect spin-offs from research (Fontes, 2001) and all except one were created by young people who had just graduated or were conducting research in the context of short time appointments, fol-lowing their postgraduate studies. They are often very small firms, but the number of employees can be misleading, since all firms resort more or less extensively to offering research and business training opportunities (from

Table 7.1   General characteristics of firms in case studies

Aspect	A	B	C	D	E	F
Age	1990	1992	1994 (1999)	1996	1998	1999
Founders	Researcher	Young graduates Research fellows	Young grad & postgrad	Young postgrad Research fellow	Young grad & postgrad	Young postgrad Research fellows
Appl. area	Agro-food	Agro-food	Agro-food Health	Health	Environ-ment	Health
No. emp. (2002)	14	12	30 (14)	25	5	5
Output	Product Services	Product Contract R&D	Tech-nology Services	Tech-nology Contract R&D [Product]	[Product]	Contract R&D
Main business	In market with product	Entering market with product	Entering market with tech-nology	In market with tech-nology Develo-ping product	Develo-ping product	Entering market with services

graduation to postdoctoral level) to young people, who complement their teams.

4.   THE STRUCTURE, COMPOSITION AND RATIONALE OF FIRMS' RELATIONSHIPS

In this section we will analyse the structure and composition of firms' relationships – with regards to their access to scientific and technological knowledge and the markets – and we will attempt to understand the motives and strategies underlying their establishment. Although, for methodological reasons, technological and market relationships will be addressed separately, it is acknowledged that, in this type of firms, market-oriented and technology-oriented relationships can be closely interrelated, with decisions and outcomes in one area often impacting on and shaping the other.

4.1 Access to Knowledge: Scientific and Technological Relationships

The first propositions address the relative importance of local and distant relationships in the access to and the development of scientific and technological knowledge. They assume that the local/national context – namely
the organization(s) from which firms spin off – is particularly relevant in this
domain at start-up, although it will very early (often from inception) need to
be complemented by relationships with foreign organizations. Another
proposition suggests that the need to expand scientific and technological
linkages will increase through time, as firms' development require other competencies or as they move into new fields, less developed locally.

This issue will be addressed in two steps. First, we will provide a general
overview of the formal research relationships established by this group of
firms; then we will address in detail the process that led to the establishment
of these relationships.

4.1.1 General overview

Table 7.2 summarizes the data on formal relationships established by the
firms in the context of research projects or contracts. It shows that, with one
exception, both foreign partners and the projects in which they participate,

Table 7.2   Formal research projects

Aspect	A	B	C	D	E	F	Total
No. projects	5	13	5	5	2	5**	35
Projects with foreign partners	*40.0%*	*61.5%*	*60.0%*	*60.0%*	*50.0%*	*100%*	*62.9%*
No. individual partners	7	41	16	13	7	12	96
Foreign partners	*14.3%*	*78.0%*	*68.8%*	*76.9%*	*71.4%*	*91.7%*	*72.9%*
Year firm creation	1990	1992	1994–99*	1996	1998	1999	–
Year first project	*1994*	*1992*	*1999*	*1997*	*2000*	*1998***	–
Projects / age	5/12	13/10	5/3	5/6	0.5	1.6	–
Partners / age	7/12	4.1	5.3	2.2	1.8	4.0	–

Notes:
* Firm was created in 1994, but biotechnology activity only started formally in 1999.
** Founders worked as consultants in parent projects; one project started before the formal
start-up.

are more frequent than exclusively national ones. If we look at the individual firms it is possible to conclude that neither the level of participation in external research projects, nor the relative importance of foreign partnerships seems to be related to age. They are basically firms, of different ages, that invest more in external research projects (B, C, F) and firms that invest less (A, D, E). Only two projects were funded by private sources. All the others were developed in the context of national or European funded government programmes.

The 35 projects analysed involved 127 cases of one-to-one collaborations. Table 7.3 presents data on partner type and location. Globally, there are more collaborations with foreign organizations (63 per cent), particularly from Europe and there is no predominance of collaborations with local organizations over other national ones, which is possibly related to the small size of the country. Research collaborations with academic organizations prevail over those with other firms because there is almost a complete absence of firms with Portuguese partners. This general pattern can be found in the majority of the individual firms, although the relative importance of collaborations with ROs versus other firms and the relative weight of local collaborations versus national ones differ among them. With respect to joint patents and publications, when they exist (since not all firms invest at these levels), they are filed/co-authored with partners from research projects. Only one patent was jointly filed with a commercial partner. The principal countries of origin of partners, apart from Portugal, are the UK, followed by France, Germany and Spain, with very few partners from outside Europe.

Quantitative data on the number of formal projects and collaborations can give an idea of the intensity of foreign connections, but it only tells a part of the story. In fact, when asked, firms said that they tended to select a smaller subset which they regarded as the key partners. These might have been around for long periods, even if this was not reflected in the participation in repeated projects and if some relationships were never formalized in projects or

Table 7.3   Total collaborations by location and type of partner

Location	Academic	Firm	Total by location
Local	25	2	27
National	20	–	20
European	37	34	71
Other foreign	7	2	9
Total by type	89	38	127
	(70%)	(30%)	(100%)

contracts. Thus, a better understanding of the nature of relationships requires an analysis of the way in which they were established and evolved.

4.1.2   The origin and evolution of scientific and technological relationships

An analysis of the sources that contributed to the formation of a firms' early knowledge base, which included the origin and background of the entrepreneurs and the set of formal and informal linkages they established during the launch process and in the early stages of firms' development, enabled us to conclude that there is a dual influence of local/national and foreign sources, although the relative weight of each factor differed from firm to firm.

Co-location with one or a set of local ROs was instrumental for the start-up stages, but for different reasons. The local ROs played one of the following roles:

1. Knowledge generated locally was the basis for the firm's formation, with technology being transferred from local ROs to the new firm, whether the entrepreneurs had directly been responsible for knowledge production (as researchers or students) or had later been involved in the later development stages (cases C and F).
2. Local ROs were the *setting* where the technology was developed, the entrepreneurs being afforded conditions to link to or integrate research teams (e.g. through pre-start-up research fellowships or joint projects) (cases A, B, D, E). But, these ROs could be more or less limited, as the sources of critical knowledge and development was largely based on entrepreneurs' own competences, obtained through their experience abroad, and on the linkages they established with foreign organizations. At the limit some entrepreneurs have initiated new lines of research in the RO (B and D).

There was usually a local RO that assisted the entrepreneurs in their initiative, either as the main source of knowledge or as an additional contributor, providing them with some conditions to develop their research activities – access to facilities and in-house competencies, scientific contacts, credibility towards funding sources and research partners. Additionally, internationally well-connected ROs would afford their young researchers or research partners access to their international networks.

With respect to the influence of knowledge originating outside the national environment, it was often at work even before the firm was created, through the entrepreneurs' previous activity. In fact, all firms had at least one founder (and sometimes also employees) with some international experience, namely through PhDs partly or totally conducted at foreign universities, or through previous participation in international projects. This early experience exposed them to knowledge developed in centres of excellence and to a variety of environments and cultures and enabled the development of personal networks, that were later useful for the firm.

All the firms – even those based on knowledge generated from world class local ROs – had established (or pursued) extensive relationships with foreign organizations during the early stages. The motives behind this early need to access foreign sources could be different:

1. When firms were launched in fields that were underdeveloped at country level, the need was obvious: the firm had to identify and establish contacts with complementary sources of knowledge and, in extreme circumstances, had to search for its main partners abroad. This entailed extensive efforts on the part of the firm. In this search the local 'parent' could be a source of credibility and eventually of resources, or even get involved in joint projects (which in some cases led to ROs' subsequent engagement in the processes of co-development of competences).
2. When the local science base was strong in the target field, this usually coincided with the extensive internationalization of local research teams. Thus, the firms had to integrate the 'parent' international scientific network in order to gain access to or participate in the development of new knowledge. For this reason, the firms' early foreign links were often established in the context of parent projects or through contacts within the parent network. Additionally, the almost complete lack of sources of more application-oriented knowledge, due to the absence of large local companies interested in biotechnology, forced all firms to search abroad for this type of competence.

We have also analysed the evolution of relationships, separating between those directed to the development of existing areas and those related to the eventual entry into new fields. Globally, the analysis revealed a growing importance of foreign relationships, associated with different levels of (dis)connection from the local environment. Concerning the motives behind the new foreign relationships, it was concluded that they were predominantly used for developing new competences, but in areas pertaining to the firm's 'core' field. Entry into new application areas or into new fields was only attempted by the older firms, which also led to the expansion of foreign relationships. These moves were often based on existing foreign partnerships or were at least triggered by the integration they afforded into wider networks, but in one case they entailed an independent search for new partners in a completely new field.

The exception to this focus on new foreign partnerships was protagonized by the older firm, whose exploratory search in a new field involved local ROs conducting world class research in that domain. Interestingly these were the same organizations whose expertise had already originated other NBFs. This calls the attention to the role still played by the local environment to these firms. The research has shown that local ROs that were critical sources of early knowledge maintain an important role towards their spin-offs, while these early supportive ROs that engaged in processes of co-development or parallel development of competencies, continue to provide some contributions. In a context of growing integration in international networks, these ROs remain the main link of these firms to their local environment and can contribute in different ways to consolidating their foreign search efforts. On the contrary, in the cases where this local background is absent or too weak, firms appear to be increasingly disconnecting from their local environment.

4.2   Market Relationships

In the case of market relationships it was proposed that firms will need to search for foreign markets during the early stages when they have limited support in that search, both in terms of resources (particularly financial) and in terms of alliances. We have looked at the firms' market strategies and have analysed their market-oriented relationships in terms of clients and partners. Notice however that the information that it was possible to collect regarding market relationships was much less systematic and detailed than that on technology-oriented relationships.[2]

Contrary to our expectations we found that the majority of firms started with activities targeting the national market and that, for some, national clients were the main source of revenue in the early years. However, a closer analysis shows that, because of the deficiencies in capital and technology markets, these firms have in fact addressed the national market with services (and in a few cases products) *that were not part of their core business*, in order to sustain themselves while developing their technology, or to raise the income necessary to start financing that development and, in some cases, also to prospect the responsiveness of the market. This strategy is frequent among NBFs (Pfirrmann, 1999), but the problem here was that, while these activities could be more or less related to the prospective core business, they almost always targeted a market that was *less demanding* in technological terms. Thus, not only would development slow down (with potential impact on competitiveness), but firms had fewer opportunities to accumulate knowledge useful for subsequent activities (Clarysse et al., 2001).

With respect to the core business, the focus was definitively international, whether foreign markets were targeted from the start-up, or firms ended up reconfiguring their strategies towards them. The latter move often involved abandoning their initial idealistic expectations concerning their role in improving the country's industrial capabilities in biotechnology and their contribution to upgrading user sectors, which were confronted with the relative indifference of local industry with regard to this technology. Only two firms continued regarding the national market as a relevant, albeit only a partial, outlet for their core business, and only one of the firms has already proved the viability of combining the national and foreign markets in the targeted niche market.

As a result of these processes, we have a group of firms whose market *orientation* is indeed international, although it is the Portuguese market that (temporarily) sustains some of them. But, with regards to the core biotechnology products, services or technologies, the market is international and firms need to search abroad for clients and partners right from the up-start. The behaviour of local firms and capital markets also means that resources and alliances to support this search are difficult to come by in the national environment.

The data limitations make it possible to only provide a generic overview of the market-oriented foreign relationships. The type of relationships firms seek to establish is associated with the type of market they target: for technology, products or services. But Table 7.4 however, shows that

Table 7.4   What type of relationships firms looked for in foreign markets

Type of relationship	A	B	C	D	E
Target market	P	P/S	T [S]	T/S [P]	S
Clients & distribution channels for products	3	3	–	3	–
Clients for technology	–	–	3	3	–
Clients for R&D services	–	3	3	–	3
Industrial partners for technology development	–	–	3	3	–
Partners for production & distribution	3	3	–	3	–
Partners for foreign expansion services	–	–	3	–	–
Technology suppliers	–	–	3	–	–
Opportunities for critical information exchange	–	–	–	3	3
Variety of client needs to shape & test product	3	3	3	–	–

T=technology; P=products; S=services; [ ]=in near future

most firms look for a range of partnerships: clients, partners for technology development or for production or for distribution. Some firms also relied on foreign market relationships either for information or as a source of variety in clients' needs (especially for those firms that also targeted the national market).

The most important market relationships are in Europe, particularly Northern European countries (namely Germany, UK and Belgium) and also Spain. Nevertheless there are some attempts to reach markets outside Europe – USA, Australia, Canada – particularly by firms targeting global niche markets. On the whole, large firms were more prevalent among clients and partners, although some firms had or were establishing partnerships with other NBFs, namely earlier partners from R&D projects.

4.3   Establishing Distant Relationships

In this section we analyse in more detail the strategies and mechanisms used by firms to identify potential partners, establish relationships and manage them. The propositions put forward regarding this process suggested differences in the establishment of technological and market relationships and argued for the importance of mediation through local scientific networks and the entrepreneurs' personal networks, during the search processes.

The research has shown that, in fact, the search for foreign partners – especially in the case of technological relationships – was frequently based on previous contacts, or at least mediated through them. In the case of technological relationships, processes of direct mediation protagonised by internationally connected local ROs, that afforded firms' admission into their international scientific communities, were already described. However, this mechanism was not available to all firms or in all circumstances and thus NBFs had often to conduct their own search for relevant sources of knowledge. In the case of market relationships, direct mediation was rarely available (with the possible exception of that afforded by EU brokerage mechanisms), given the absence of local firms acting as channels to foreign markets and NBFs had to rely almost exclusively on their own search activities.

In both cases the entrepreneurs' personal networks, derived from their previous international experience, played an important part in these 'own search' processes, although they were more useful in the technological area, given the prevailing backgrounds. But some firms still considered that part of their search had gone completely unsupported, particularly in early stages, suggesting that in some circumstances not even personal networks could be mobilized to access the relevant resources and partners. This situation tended to change with time, as firms gained international experience, reputation and a wider range of contacts that assisted their subsequent search efforts, enabling them to build upon existing relations, or making them visible to potentially interested partners.

Given the importance assumed by the firms' 'own search', it is relevant to understand how it took place. Unsupported search for scientific and technological competencies was more frequently conducted towards ROs. Identification and first contacts (usually between individual scientists) were generally easier. But, unless entrepreneurs had been previously been involved in knowledge co-production with these scientists or teams, or had a high scientific record in the field or some competence that the potential partners regarded as immediately interesting, attainment of status and development of trust could be a slow processes. In three of the cases that fell within this mode, relations with the organizations that are now key technological partners progressed through a similar staged process, starting with formal contract research funded by the firm and progressing, through growing levels of involvement, as the firms' competencies developed and conditions for effective collaboration were attained. Indeed some firms also used this slow route as an opportunity for further capability building. This process involved the exchange of people (namely stays in the partner facility) which enabled the development of closer personal relationships. An alternative process entailed prolonged informal exchanges that enabled mutual awareness of skills and interests, leading to an eventual identification of joint interests. Such links could stay informal and relatively unfocused until some opportunity was identified.

Unsupported search for market relationships showed some differences and, on the whole, was regarded by firms as more difficult to pursue. It was a long term systematic process which involved: identification of opportunities; search for targets; achieving contact; credibilization of the company; and negotiation. The first stages – identification, search and contact – could be much longer than in technology relationships, with firms mentioning the difficulties of achieving contact, particularly with large firms. Credibility checking could be very stringent and the negotiation processes very protracted, especially in early stages, when the firms do not have a reputation and particularly for firms selling technologies or R&D services. As firms became better known and 'word of mouth' was at work, this early suspicion was likely to lessen a little, but there was always a competence scrutiny procedure. There were some differences between large firms and other NBFs in this process, with the first (which were the among the firms' most important clients and potential partners) being much harder to access and to reach agreements with. Relationships with other NBFs often emerged from R&D projects or from personal networks and, either for this reason, or because entrepreneurs spoke a similar language, they were easier to establish.

Personal networks played different roles in these processes. In market relationships they served mostly as facilitators, that is firms used their contacts to achieve access and to provide first references, although these had to be demonstrated during negotiation. In the early stages they were basically social networks (such as, ex-colleagues working in foreign firms) and it was only later that did they started being composed of previous clients or market partners. With respect to technology access, members of the personal networks, (often ex-research partners, colleagues or supervisors) could themselves be the target for collaboration, provide access to their scientific networks, or act as credibility enhancers. The latter role was namely performed by reputed scientists whom the firms enlisted as an informal 'advisory board'. Informal linkages with reputed scientists were rarely used as references for business, but formal research partnerships were used for that purpose.

With respect to the mechanisms used in the search process, it was concluded that while ICT is widely used in the field, both for business and for research, face to face contacts remain critical. ICT can partly assist the early identification of opportunities, can support the activities, particularly when they have been formalized and are ongoing and can assist in nurturing personal networks. But the effective establishment of relationships requires face to face contacts at some point, negotiation processes require frequent interaction and the development of technology relationships may require periods of temporary co-location. Direct contacts are also necessary for ongoing partnerships, even if only occasionally, to guarantee periodic reassessing of issues and to maintain the relationships in good shape. Finally, attendance at key international events that bring together the main scientific and/or industrial actors in a given area, can also be an important source of information about opportunities and a fruitful means of making new contacts.

These results are not exclusive to out-cluster NBFs. But their most significant implication for these firms is that, because a substantial part of their contacts will be distant, out-cluster entrepreneurs will need to constantly travel great distances in order to guarantee a level of integration at least close to those who have a more substantial part of their partners', clients' and personal networks nearby. Additionally, cultural differences will be more critical for these entrepreneurs and country-of-origin effects may be at work, making negotiation processes slower and still increasing the costs and difficulties of reaching agreements. This will entail a much greater financial cost, and personal effort than is required by similar firms located in clusters Moreover, members of out-cluster NBIs also require particularly good relational skills.

5. DISTANT NETWORKING STRATEGIES

The analysis conducted in the previous sections enabled us to go through the initial propositions regarding the adoption of specific strategies by biotechnology firms operating out-cluster and permitted an in-depth understanding of the conditions underlying them. While generically confirming the propositions, the analysis of the particular cases permitted us to identify some variety regarding the relative relevance of the national/ international environment, as well as diverse forms of addressing the general conditions all firms faced. It is therefore possible to advance a first characterization of what we have labelled 'distant networking strategies':

1. Relevance of co-location to a particular RO or set of ROs in the processes that lead to the creation and early development of the firm. *But different weights of main RO inputs: knowledge/capacity to assist development*;
2. Need to resort to foreign relationships at early stages, in order to complement the national knowledge base and the resources available locally. *But different levels of national/foreign contribution, depending on strength of national knowledge base; and different levels of mediation in search processes*;
3. Critical importance of foreign markets and of foreign market relationships for the commercialization of core technologies/products. *Importance of national market in early years, as source of income while developing the core business, but only for less sophisticated services or products*;
4. Unsupported search for foreign clients and market partners, given weakness of industrial structure in relevant areas and deficiencies of national capital markets. *Although capacity to conduct this search differed, according to founders' foreign experience and type of personal networks*;
5. Intensity of purposive/planned interactions, involving frequent face to face contacts, in addition to extensive use of ICT, hence requiring high relational capacities and constant travel (with associated costs);
6. Influence of entrepreneurs' (and employees') international background, experience and contacts in technology and market internationalization processes;
7. Potentially negative impact of 'country-of-origin' effects.

Notwithstanding these common features, it is possible to devise two major types of strategic approaches to building up foreign relationships, which are basically influenced by the presence and the quality of the local knowledge base in relevant fields and by the degree of integration of local ROs in international scientific networks. In fact, the majority of firms had, from the start-up, perceived the foreign market as an important outlet for their business, be it complementary or exclusive, and they were mostly unsupported in their search in this area. Therefore, the *conditions* in which firms approached foreign market relationships were relatively similar, even if the *modes* could be different. On the contrary, firms differed in terms of: (1) the relative *need* for knowledge originating from foreign sources; (2) the *conditions* in which they searched for these sources and their ability to gain access to and establish relationships with them, as well as the capacity to absorb and use the knowledge thus acquired. The main source of such variance was the *strength of the national science base*.

Two different patterns were thus identified in the establishment of foreign technological relationships:

Pattern 1: *Mediated integration* – Based on a strong national science base, embodied in the 'parent' ROs, who also have a good integration in international scientific networks.

Pattern 2: *Exploratory integration* – Based on weaker or still developing national science base and on limited connections with international research, but associated with the local ROs interests, and the assistance of, entrepreneurs' efforts.

The main features of *mediated integration* are:

1. The national science base, characterized by high quality and consolidated research conducted in one or a set of ROs. The ROs have a significant bearing on the decision to establish firms. And in the early stages they are also one of the firm's main sources of knowledge.
2. The production of knowledge usually takes place in the context of international scientific networks, in which the parent organization(s) play a relevant part. Thus firms need to access complementary knowledge that is distributed in the network. Particularly they need to access and participate in the production of more application-oriented knowledge (that is absent locally), collaborating with foreign firms for this purpose.
3. ROs' willingness to provide access to their network (when entrepreneurs are not already part of it) enables a firm's participation in common research projects as well as less formal exchanges. This mediation eases entry into research communities where access might be difficult for newcomers. Integration in the community and participation in technology development facilitate the access to more tacit forms of knowledge, favouring absorption and may also generate new opportunities.
4. Through time the firm and its scientists reduce dependence on the parent for access and may become network members in their own right

and pursue with further activities within both the specific network and other connected networks.

5. Contacts with technological partners may even progress to market-oriented relationships or be of use in the search for such relationships.
6. Finally, if the parent RO (or other local ROs with whom the firm collaborates) is scientifically strong and pursues high quality research, it may remain an important source of knowledge, credibility and contacts.

The main features of *exploratory integration* are:

1. National science base is less strong, or still being developed (sometimes also through the pioneering activities of the new firm), or does not have an application-oriented nature. ROs are interested in and supportive of entrepreneurs' activities, smooth access to facilities and existing competencies and provide institutional credibility. But their effective knowledge contribution is definitively lower that in Pattern 1, the development process is usually in a less advanced stage and the need for complementary knowledge is much wider.
2. Parent organizations may have some scientific relationships in the field, but their degree of interaction with international research conducted in the area will also be much lower. Therefore, they may still provide some contacts and offer institutional credibility, that assist search for foreign relationships, or may just provide a setting where entrepreneurs have better conditions to develop their own competences, to pursue their search activities and to start building upon the results of that search.
3. Access to complementary knowledge through foreign relationships depends much more on firms' efforts. The more frequent absence of direct mediation by reputable members of existing networks inevitably entails slower processes, not only in terms of identification of suitable partners, but particularly in terms of acceptance by them, development of trust and an eventual integration into 'research communities'.
4. Personal networks are instrumental: entrepreneurs with previous international background may build on previous co-development experiences to launch new relationships or, at least, benefit from the indirect mediation of well positioned ex-supervisors, professors and colleagues. On the other hand, if the firm is able to establish good relationships with a few key actors, these may become a sort of gateway to the wider network, in a fashion not dissimilar to that performed by a local 'parent'. But the effort is greater, dead ends more frequent and success less certain.

5.   When the firm is successful in its efforts, it may start benefiting from the advantages of becoming a network member in its own right, as already described in Pattern 1.

Distant networking strategies were a basic feature of Portuguese NBFs' behaviour. The fact that firms were able to establish and manage this specific form of knowledge acquisition and market access shows that geographical distance may not be a deterrent to firms' development – even if they face some specific difficulties – providing that they are able to profit from other forms of proximity, devising the adequate strategies.

6.   CONCLUSION

The analysis of a group of biotechnology firms created in Portugal has provided some evidence regarding the conditions in which these firms are formed and developed outside biotechnology clusters; locations where knowledge accumulation is lower and some of the critical actors are missing. It was argued that while clustering is important for the evolution of this sector, biotechnology also presents some features – namely the international nature of scientific production and markets – that may facilitate firm development outside them. But, it was also pointed out that the firms' ability to survive and grow in these environments cannot be regarded as evidence that location is immaterial – indeed, the small number of firms that manage to materialize is evidence of the contrary! Rather, it means that these firms have been able to devise strategies to overcome some of the relative disadvantages of their location, enabling them to access and integrate nonlocal networks, to draw creatively from a combination of local and distant relationships and to manage this specific form of knowledge acquisition and business development.

The results of the empirical research confirm that, for the firms studied, distant relationships are a critical source of competencies and resources from the start-up and that their relevance increases through time. Less systematic evidence from younger firms (also being followed up, but not included in this more in-depth analysis) point in the same direction. The *early* need to access and integrate distant (bio)technological networks differentiates these firms from those located in more knowledge intensive environments (Lemarié et al., 2001).

More specifically, the research enabled us to characterize a 'distant networking strategy', as follows. Firm formation decisions are associated with the presence of local sources of scientific knowledge, with which close relationships are established; but firms will also develop, from inception, a set of transnational connections, based on the entrepreneurs' own networks or accessed through local research organizations. Firms draw, at least in early stages, upon a combination of local and nonlocal sources to access scientific and technological knowledge, but they tend to search externally for markets and market-related relationships. Connections to external networks expand and become increasingly important along the firms' life cycle, as they progress towards the commercialization stages and/or need to broaden or renew their knowledge base.

With respect to establishment of foreign relationships, mediation through local scientific partners or through personal networks is key, although more frequently available for technological than for market relationships, making the latter generally more complex to establish. With respect to the problem of long distance transmission of more tacit or 'excludable' types of know-ledge, it can be concluded that a form of 'epistemic proximity' to relevant scientific communities was achieved by firms through integration in the 'parent' scientific networks, or through previous co-development experiences with members of entrepreneurs' personal networks and, at later stages, through extensive investment in temporary location of people in foreign centres of excellence. Additionally the fact that firms were looking for knowledge that was not too distant from their own knowledge bases – rather contributed to developing or expanding it – facilitated this process, configuring situations of 'technological proximity at geographical distance'.

With respect to the mechanisms used, it was found that while ICT means are important to identify and make first contact with partners and to maintain already ongoing relationships, face to face contacts remain critical for the effective establishment of relationships – especially when the process is not mediated or in the case of market relationships – and temporary co-location is essential for technology development. For these reasons, there is a need for constant travel to establish or renew contacts, attend events or relevant meetings, pursue with negotiations or coordinate ongoing projects, as well as for periodical longer stays for co-development purposes. This requirement has high costs, both in financial and personal terms. Additionally, firms experience the combined impact of geographical distance and cultural differences on the speed and smoothness of negotiation processes and on the development of trust. All this may require particularly good relational skills on the part of entrepreneurs.

In conclusion, operating at a distance from the main biotechnology centres where potential partners and clients locate is viable, but it has influence upon NBFs' behaviour, raising particular problems and requiring specific strategies. Distance is more significant in the early years, when firms are still building their relationships and lack the credibility afforded by reputation or the mediation provided by a wider network of contacts. With time they tend to become more integrated in foreign networks and learn to deal with the difficulties of distance. However, the additional costs and management complexity may lead some firms to question their location . . . unless they retain some of their early 'missionary' vision of a role in the development of the Portuguese biotechnology industry.

NOTES

1. *Methodological Appendix*
   The collection of hard data about relationships involved searches in a variety of national and foreign databases for R&D projects and patents and the consultation of firms' web pages, as well as other documentation available on them. The information obtained was subsequently checked with the firms.
   The interviews took place during the second half of 2002 and early 2003. The following people were interviewed, at least once and in a number of cases twice:
   Firm A – Founder; R&D Director
   Firm B – Founder
   Firm C – Founder; entrepreneur joining later specifically for biotechnology area
   Firm D – Founder (CEO); entrepreneur joining later (COO)
   Firm E – Founder responsible for commercial area; founder responsible for R&D
   Firm F – Founder
   Additional written information was supplied by some firms before and after the interviews. In some cases, data analysis and interpretation of results required further discussion with the interviewees, conducted over the telephone or by email.
2. Because often firms had not yet introduced their products in the market or were in early stages of commercialization, they could only describe their attempts at identifying and contacting potential partners and clients. Also, given the secrecy frequently involved in market transactions, firms were often reluctant to mention the name of clients and the type of business involved. In these cases we have tried to elicit, at least, the countries of origin and the basic characteristics (size, sector) of their principal clients and of potential clients.

REFERENCES

Allansdottir, A., A. Bonaccorsi, A. Gambardella, M. Mariani, L. Orsenigo, F. Pammolli and M. Riccaboni (2002), '*Innovation and Competitiveness in the European Biotechnology Industry*', Enterprise Directorate General, enterprise papers no. 7, European Comission.

Amin, A. and P. Cohendet (2003), 'Geographies of knowledge formation in firms', Danish Research Unit for Industrial Dynamics (DRUID) Summer Conference 2003 on Creating, Sharing and Transferring Knowledge, Copenhagen, 12–14 June.

Arora, A. and A. Gambardella (1994), 'Evaluating technological information and utilizing it: scientific knowledge; technological capability and external linkages in biotechnology', *Journal of Economic Behavior and Organization*, **24** (1), 91–114.

Arora, A. A. Fosfuri and A. Gambardella (2001), 'Markets for technology and their implications for corporate strategy', *Industrial and Corporate Change*, **10** (2), 419–51.

Arundel, A. and A. Geuna (2001), 'Does proximity matter for knowledge transfer from public institutes to firms?', SPRU electronic paper series, no. 73.

Audretsch, D. B. and P. E. Stephan (1996), 'Company–scientist locational links: the case of biotechnology', *American Economic Review*, **86** (3), 641–52.

Barbanti, P., A. Gambardella and L. Orsenigo (1999), 'The evolution of collaborative relationships among firms in biotechnology', *International Journal of Biotechnology*, **1**, 10–29.

Bathelt, H., A. Malmberg and P. Maskell (2002), 'Clusters and Knowledge: Local Buzz, Global Pipelines and the Process of Knowledge Creation', DRUID working paper no. 02–12.

Breschi, S. and F. Lissoni (2001), 'Knowedge spillovers and local innovation systems: a critical survey', *Industrial and Corporate Change*, **10** (4), 975–1005.

Clarysse, B., A. Heirman and J. J. Degroof (2001), 'An institutional and resource-based explanation of growth patterns of research-based spin-offs in Europe', *STI Review*, **26** (1), 75–96.

Cohen, W. M. and D. A. Levinthal (1990), 'Absorptive capacity: a new perspective on learning and innovation', *Administrative Science Quarterly*, **35** (1), 128–52.

Cooke, P. (2001), 'Biotechnology clusters in the U.K.: lessons from localisation in the commercialisation of science', *Small Business Economics*, **17** (1/2), 43–59.

Cooke, P. (2002), 'Towards regional science policy? The rationale from biosciences', Conference Rethinking Science Policy, University of Sussex, 21–23 March.

Coviello, N. E. and H. J. Munro (1995), 'Growing the entrepreneurial firm: networking for international markets', *European Journal of Marketing*, **29** (7), 49–61.

Crick, D. and M. Jones (2000), 'Small high-technology firms and international high-technology markets', *Journal of International Marketing*, **8** (2), 63–85.

Echeverri-Carroll, E. and W. Brennan (1999), 'Are innovation networks bounded by local proximity?' in M. M. Fischer, L. Suarez-Villa and M. Steiner (eds), *Innovation, Networks and Localities*, Berlin: Springer Verlag, pp. 28–47.

Feldman, M. (1999), 'The new economics of innovation, spillovers and agglomeration: a review of empirical studies', *Economics of Innovation and New Technology*, **8** (1/2), 5–25.

Felsenstein, D. (2001), 'New spatial agglomerations of technological activity – anchors or enclaves? Some evidence from Tel Aviv', presentation to annual conference of IGU Commission on the Dynamics of Economic Spaces, Turin, Italy, 10–15 June.

Fontes, M. (2001), 'Biotechnology entrepreneurs and technology transfer in an intermediate economy', *Technological Forecasting and Social Change*, **66** (1), 59–74.

Fontes, M. and M. Pádua (2002), 'The impact of biotechnology pervasivenss and user heterogeneity on the organisation of public sector research', *Technology Analysis and Strategic Management*, **14** (4), 419–41.

Gallaud, D. and A. Torre (2001), 'Les Réseaux d'Innovation sont-ils Localisés? Proximité et Diffusion des Connaissances (Le Cas des PME de l'Agbiotec)', Third Congress on Proximity 'New Growth and Territories', Paris, 13–14 December.

Lemarié, S., V. Mangematin and A. Torre (2001), 'Is the creation and development of biotech SMEs localised? Conclusions drawn from the French case', *Small Business Economics*, **17** (1/2), 61–76.

Mangematin, V., S. Lemarié, J. P. Boissin, D. Catherine, F. Corolleur, R. Coronini and M. Trommetter (2002), 'Development of SMEs and heterogeneity of trajectories: the case of biotechnology in France', *Research Policy*, **32** (4), 621–38.

McKelvey, M., H. Alm and M. Riccaboni (2003), 'Does co-location matter for formal knowledge collaboration in the Swedish biotechnology-pharmaceutical sector e.m.', *Research Policy*, **32** (3), 483–501.

McKelvey, M. and L. Orsenigo (2001), 'European pharmaceuticals as a sectoral innovation system: performance and national selection environments', second European Meeting of Applied Evolutionary Economics (EMAEE), 13–15 September.

Orsenigo, L., F. Pammolli and M. Riccaboni (2001), 'Technological change and network dynamics: lessons from the pharmaceutical industry', *Research Policy*, **30** (3), 485–508.

Owen Smith, J., M. Riccaboni, F. Pammolli and W. Powell (2002), 'A comparison of US and European university–industry relations in the life sciences', *Management Science*, **48** (1), 23–43.

Pfirrmann, O. (1999), 'Neither soft nor hard – pattern of development of new technology based firms in biotechnology', *Technovation*, **19** (11), 651–59.

Powell, W., K. W. Koput and L. Smith-Doerr (1996), 'Interorganizational collaboration and the locus of innovation: networks of learning in biotechnology', *Administrative Science Quarterly*, **41** (1), 116–45.

Pyka, A. and P. Saviotti (2000), 'Innovation networks in the biotechnology-based sectors', Eighth International Joseph A. Schumpeter Society Conference Manchester, 28 June–1 July.

Queré, M. and P. P. Saviotti (2002), 'Knowledge dynamics and the organisation of the life sciences industries', paper prepared for the DRUID Summer Conference on 'Industrial Dynamics in the New and Old Economy – Who is Embracing Whom?' Copenhagen, 6–8 June.

Rees, K. (2001), 'Collaboration, innovation and regional networks: evidence from the medical biotechnology industry of Greater Vancouver', Department of Geography, working paper, University of Wales, Swansea.

Roberts, J. (2000), 'From know-how to show-how? Questioning the role of information and communication technologies in knowledge transfer', *Technology Analysis and Strategic Management*, **12** (4), 429–43.

Rosenkopf, L. and P. Almeida (2003), 'Overcoming local search through alliances and mobility', *Management Science*, **49** (6), 751–66.

Roth, M. S. and J. B. Romeo (1992), 'Matching product category and country image perceptions: a framework for managing country-of-origin effects', *Journal of International Business Studies*, **23** (3), 477–97.

Saxenian, A., and J.-Y. Hsu (2001), 'The Silicon Valley-Hsinchu connection: technical communities and industrial upgrading', *Industrial and Corporate Change*, **10** (4), 893–920.

Steinmueller, E. (2000), 'Does information and communication technologies facilitate "codification" of knowledge?', *Industrial and Corporate Change*, **9** (2), 361–76.

Stuart, T. and O. Sorenson (2003), 'The geography of opportunity: spatial heterogeneity in founding rates and the performance of biotechnology firms', *Research Policy*, **32** (2), 229–53.

Swann, P. and M. Prevezer (1996), 'A comparison of the dynamic of industrial clustering in computing and biotechnology', *Research Policy*, **25** (7), 1139–57.

Walsh, J. and T. Bayma (1996), 'The virtual college: computer-mediated communication and scientific work', *Information Society*, **12**, 343–63.

Zeller, C. (2001), 'Clustering biotech: a recipe for success? Spatial patterns of growth of biotechnology in Munich, Rhineland and Hamburg', *Small Business Economics*, **17** (1/2), 123–44.

Zucker, L., M. Darby and M. Brewer (1998), 'Intellectual human capital and the birth of US biotechnology enterprises', *American Economic Review*, **88** (1), 290–306.

8. New science and old industries: adoption of biotechnology in European food companies

Finn Valentin and Rasmus Lund Jensen[1]

1. INTRODUCTION

Through the 20th century the life sciences became an important source of innovation and economic development, and that importance is expected to grow further over the next decades. At the same time, its discovery process and further linkages to technologies and applications have come to depend on complicated (inter)organizational forms which in turn are quite sensitive to institutional influences and regulation (Cockburn et al., 1999).

Consequently, interdependencies between these organizational forms and the economic performance of life science-based industries have attracted interest since the onset of the biotech revolution. This interest increased as the US model for biotech competitiveness through the 1990s became idealized as the model against which other countries could be benchmarked. But this idealization needs scrutiny. We need to better understand if the success of the US model is specific to particular areas – or stages – of biotechnology. Will the infusion of biotechnology into agriculture and foods require other models? And will different organizational and institutional forms prove equally successful as other countries move biotechnology into new fields of application. To learn from the US experience we must see it in comparative perspective (Chesbrough, 2001; Lynskey, 2001).

Everywhere biotechnologies induce distributed forms of innovations (Coombs and Metcalfe, 2000), involving networks of collaboration between large firms and outside partners (Liebeskind et al., 1996; Powell, 1998; Sharp and Senker, 1999). The formation of more than a thousand new Dedicated Biotechnology Firms (DBFs) is emphasized as a crucial component in the US model for biotech success. Their emergence is interpreted as a classical case of Schumpeterian industrial transformation caused by the technological discontinuity of the biotech revolution. But may Schumpeterian transformation also take place in very different organizational forms?

1.1   Issues and Objectives

We examine in this chapter a case where the discontinuity of biotechnology generates a type of distributed innovation very different from the US 'model'. We focus on a quite narrow field of food technology in which incumbents are not disadvantaged by discontinuities. Although highly distributed forms of innovation emerge from 1980 onwards, incumbents introduce virtually all innovations in this field. Equivalents of DBFs fail to emerge as separate units on markets for R&D. But at the same time Public Research Organizations (PROs) contribute significantly to distributed R&D, and to some extent they also take on the role of economic actors.

Analysing this case the chapter has three interrelated objectives:

1. *Descriptive objective*: Using patent data to build a comprehensive and systematic description of the way a scientific discontinuity shapes the emergence of a distributed organization of innovation and its subsequent evolution.
2. *Theoretical objective*: Contributing to an explanation of the organizational characteristics of this distributed innovation. We submit that characteristics of R&D problem processing derived from Simon's theory of complex problems contribute substantially to this explanation. A second objective of the chapter is to introduce this theoretical derivation and to bring out its implications for the effects of technological discontinuities on the organizational forms of distributed innovation.
3. *Methodological objective*: To appreciate characteristics of innovation problem solving we must understand the issues addressed in R&D. A classic dilemma in innovation research has been the restricted possibilities for characterizing large quantities of R&D activities in terms of their content, particularly their cognitive characteristics. Only a few R&D parameters lend themselves to immediate quantification, such as input and output measures of R&D, e.g. costs, patent statistics, etc. (Freeman and Soete, 1997; Grupp, 1998). These parameters, however, have limitations when it comes to characterizing the content of R&D and its resultant technologies. Richer insights into the latter require qualitative data, producing a trade-off in research designs between quantity versus depth and richness. Innovation research, for this reason, has a considerable appetite for methodologies and tools alleviating precisely this trade-off. In this spirit, the chapter tries out novel data mining tools to bring out dimensions in the text sections of patents. Characterizing content dimensions of patented biotech products and processes offers new ways of studying the agenda in large quantities of R&D projects.

We take as our case the specific field of food science and technology that utilizes Lactic Acid Bacteria (LAB). This family of microorganisms is used widely in existing food product and process technologies, and also has implications for the emerging partial fusion of food, neutraceuticals and pharmaceuticals (the role of LAB in food technologies is summarized in Appendix I). LAB appears to have been quite intensively targeted with the tools of biotechnology as they have migrated into food science from their origin in the pharma-related discovery chain. Consequently, LAB-related research and innovations offer an attractive and well delimited window on the exploitation of the new biotech science regime in food R&D. The 180 biotech-related LAB patents claimed until the year 2000 provide rich information on that exploitation, and they are the key source for the data analysed below.

The chapter is structured as follows. We first review and discuss the literature on discontinuities and distributed innovation, and relate these phenomena to a conceptual framework on R&D problem processing, derived from Simon's theory on complex problems. A short section presents methodology, primarily by guiding the reader to appendices where its specific components are explained. The two main sections first relate the R&D issues of food to the evolution of biotechnology and examine cognitive characteristics in LAB biotech R&D, its main themes and their development over the past decade. Next we identify main actors in LAB biotechnology along with their roles in its distributed forms of innovation. Their R&D profiles are identified and related to differential advantages in innovation problem definition and problem solving. The two final sections summarize results and discuss implications.

2. DISCONTINUITIES AND DISTRIBUTED INNOVATION

An influential strain in the literature on technological discontinuities links their implications to destructive effects on incumbent firms (Chesbrough, 2001). 'Competence enhancing' and 'competence destroying' consequences for firms arose as an important distinction from the studies of Tushman and Anderson of the 1980s (Anderson and Tushman, 1991;Tushman and Anderson, 1986). It defined key issues for the subsequent research agenda, including studies of the extent to which destructive effects are amenable to managerial action (Henderson and Clark, 1990), and it examined contingent cognitive and organizational conditions for such alleviation (Burgelman, 1994; Henderson, 1993).

As seminal contributions to innovation research, these studies also render the scarcity of studies addressing the twin issue of 'competence enhancement' all the more conspicuous. Technologies may be affected by substitutive or by complementary discontinuities (Ehrnberg and Sjöberg, 1995) with quite dissimilar consequences for industry competition. Substitution often gives entrants direct access to competitive positions at least in parts of the industry. The key issue of complementary discontinuities, on the other hand, is how apt companies are at exploiting a set of opportunities that in principle becomes available to their entire industry.

To exploit these opportunities faster, companies carry out their innovations by collaborating with outside partners, from whom they learn, transfer or in-source components of the new knowledge. This interorganizational coordination has been referred to as *distributed innovation* (Coombs and Metcalfe, 2000; Smith, 2001). However, it is an option only in fields lending themselves to decomposition of innovation-related tasks. This contingency was theorized in Simon's distinction between types of complex problems (Simon, 1996), only some of which may be partitioned into smaller tasks to be addressed more effectively by separate organizational units. Other complex problems have interdependencies between their constituent components preventing them from being meaningfully considered separately.

Applied to R&D problems, Simon's argument on decomposability rationalizes why sectors and technologies differ in the way they give rise to distributed innovation. High decomposability allows division of innovative tasks in which actors may then build specialized capabilities, and thus address selected components of the R&D process more effectively than do integrated innovators (Bresnahan and Trajtenberg, 1995). In response, large R&D integrators reduce their own R&D targeted at such components, effectively accepting a gradual contraction of their competitive knowledge base; or perhaps they compensate by building stronger capabilities in manufacturing or marketing instead. These economies of specialization help explain the emergence over the past two decades of a number of specialties in the pharmaceutical discovery process, particularly the emergence of more than a thousand new DBFs (Arora et al., 2001).

Distributed innovation induced by high R&D decomposability involves not only new specialized firms but also PROs, and the latter may operate in quite different capacities. PROs may contribute to problem solving in R&D consortia orchestrated by corporate lead partners, who also appropriate resultant technologies. But in other cases PROs take on the role of economic actors. They become 'quasi-firms' in the sense of initiating and orchestrating interorganizational R&D projects and being assigned resultant patents, effectively making them key appropriators of subsequent licensing arrangements (Mowery et al., 2001).

This variability in the organization of distributed innovation has been explained as an effect of strategies by which large companies build different types of external linkages to pursue different requisite goals. For example, they pursue early discoveries through DBF partnerships while using university collaborations to gain familiarity with new scientific knowledge (Gambardella, 1995).

Explaining organizational variability in distributed innovation as the effect of multiple strategies of large firms directs attention to what basis large firms would have for shaping distributed forms of innovation according to their own strategic preferences. This inquiry becomes all the more pertinent in light of the argument that large R&D integrators must adjust the boundaries of their internal R&D in response to specialization economies that are largely beyond their strategic control.

Under what conditions then are large R&D integrators in a capacity of orchestrating distributed innovation according to their own strategic interests? And when must they share that capacity with other types of actors? Multiple lines of attack are required to answer these questions exhaustively. The approach proposed in the next section is intended to theorize merely one of the dimensions that must be taken into account, while subsequent sections of the chapter will demonstrate the relevance of this one dimension to empirical analysis.

3.   DECOMPOSING PROBLEM PROCESSING OF INNOVATIONS

The argument builds on Simon's concept of problem decomposability, shown above to be at the root of most subsequent theorizing on distributed innovation. We submit that new implications of this core concept emerge if problem processing is further specified into the two dimensions of *definition* and *solution*:

1. *Problem definition* involves identification of needs and targets for inventive efforts, including insights into likely payoffs from the successful pursuit of different potential targets.
2. *Problem solving* involves building an understanding of the issue at hand, deliberation of solutions based on invention and/or combinatorial search, test and validation of results.

Although Simon does not differentiate between these two dimensions, his discussion of decomposability pertains to problem solving only. It may be extended, however, to cover both dimensions, allowing for the possibility that the two dimensions have different levels of decomposability. Figure 8.1

		Dimensions of problem processing	
		Definition	Solution
Decompose-ability of innovation processes	Low	1	3
	High	2	4

Figure 8.1   Decomposability of separate dimensions of problem processing

brings out the point that decomposable problem solving may have been preceded by a nondecomposable problem definition.

Decomposability of *problem definition* refers to *initiation* of innovative processes, and it concerns the extent to which their instigation requires an integrated view of opportunities for and utility of the prospective innovation. Problem definition may involve combined considerations of, for example, process–product characteristics and/or consumer insights. It depends on access to, observation of, and appreciation of anomalies and on an assessment of opportunities in terms of the improvements they may bring about. It has low decomposability when requiring a *confluence of different sources of information and knowledge* that will fail to suggest relevant novelties when considered separately.

Problem identification may have low decomposability even when its constituent flows of information are *highly codified*, as long as problems of potential value may be extracted only from configurations of information. The ability to assemble available information in such configurations and to extract interesting problem identification from them will rest on *local* knowledge or heuristics, i.e. the quality referred to by Eliasson as their 'economic competence' (Eliasson, 2000). The cognitive ordering produced by such local effects may turn the firm into a valuable point of confluence for flows of information which otherwise fail to offer opportunities. In this sense, they benefit from the 'economics of strategic opportunity' as this idea has recently been theorized in Denrell et al. (2003).

The point that nondecomposability (type cell 1 in Figure 8.1) appears independently of cognitive attributes of constituent single flows of information is emphasized here, since the empirical case of food technology refers to confluence of scientific findings and other types of well articulated industrial knowledge. Incumbents, we shall argue, derive their innovation advantages not from cognitive attributes of the information they process but from nondecomposability of problem identification as specified here.[2]

For some areas of innovation the distinction between the two dimensions of problem processing is superfluous because they have the same level of decomposability. As an example, on the basis of strong interdependencies in product and process technologies Bonaccorsi et al. (2001) identify a specific type of firm, which they refer to as an 'integrated system companies'. Virtually all issues of technological development in these firms – both problem definition and solution – defy decomposition, with innovation problem processing clearly conforming to a lateral combination of cells 1 and 3 in Figure 8.1. The limited possibilities for distributed innovation growing out of these conditions have pros and cons for integrated system companies: on the one hand, they are rarely put on the defensive by entrants specializing in specific parts of their innovation tasks; and on the other hand, public research or other outside partners may contribute to problem solving only at high costs of coordination (Meyer-Krahmer and Schmoch, 1998).

At the other end of the spectrum we find innovations based on *high* decomposability of *both* dimensions (lateral combination of cells 2 and 4). Given adequate specialization economies organizations may focus on specific partitions of innovation tasks. Under these conditions, divisions of innovative tasks tend to take on clearly distributed forms. The disintegration of pharma-related discovery witnessed over the past two decades fits well with this version (Cockburn et al., 1999).

While the literature, in other words, has considered effects on distributed innovation derived from both types of lateral combinations in Figure 8.1, diagonal combinations so far have not been taken into account. We shall consider only the diagonal reflected in the empirical findings presented below, i.e. the 1–4 combination.

In this combination high decomposability of problem *solving* allows innovating firms to have important parts of their innovation problems answered through specialized skills or experiences of outside partners. The latter, at the same time, would be barred from conceiving substantial innovations of their own because they do not process problems and opportunities in the type of *configurations* that give rise to interesting definition of innovation problems. In this way, the combination of cells 1 and 4 offers a favourable set-up for companies with strong R&D integration seeking to alleviate effects of discontinuities through aggressive utilization of distributed innovation. First, they are advantaged by being positioned in *the point of confluence* of critical flows of opportunities that allows them repeated extractions of valuable problem definitions. Second, once defined, these problems lend themselves to partition-friendly solution in which R&D specializations of outside partners are brought together in effective distributed innovation.

The main hypothesis of this chapter is that these advantages and their underlying levels of decomposability of problem definition and solutions largely shape distributed innovation in food biotechnology. Only few actors carry out activities making them points of confluence of diverse critical flows of information. The more such actors also carry out their own R&D, the more able they will be to extract interesting problem definitions from these points of confluence, and the better they will be positioned to orchestrate distributed innovation and make themselves key beneficiaries of its problem solving effectiveness. Actors that are not positioned at such specific points may have critical problem solving skills that will make them indispensable partners in distributed innovation, the agenda for which, however, they will not have defined.

4.   DATA AND METHODOLOGY

Patents provide most of the data analysed below. An account of patent search and processing procedures is offered in Appendix II. A total of 3425 food biotech patent families were identified, of which 180 focus on LAB technologies and applications. We use various methodologies to extract and examine information from these patents, primarily text data mining tools. The application of these methodologies is introduced in their specific sections of the chapter.

5.   BIOTECHNOLOGY IN FOOD R&D

Focusing on the *cognitive dimension* of food biotechnology, this section examines its origin in the broader revolution of molecular biology. Main trends in its evolution over the past 30 years are identified, and R&D of LAB biotechnology is analysed in terms of its key themes and their level of decomposability.

5.1   Development of Biotechnology in Food R&D

Building on the science of molecular biology and genetics accumulated over the 1950s–60s, several interrelated discoveries and inventions provided the breakthrough in biotechnology in the 1970s. This included the discovery of reverse transcriptase (1970) and the first recombinant plasmids (1973–74). The second half of the decade saw the development of cloning, of genetic libraries, and of DNA sequencing. Commercial results remained sporadic, but included the invention of genetically engineered insulin (1978) and humane growth hormone (1979). In 1982 the fundamental techniques of genetic engineering were collected and presented in *Molecular Cloning:*

a Laboratory Manual (Judson, 1979; Morange, 1998). A number of complementary technologies began to align into a coherent, effective set of technologies. High throughput screening techniques allowed the first sequencing
of entire genomes (including that of lactobacillus). Bioinformatic tools in
the form of DNA chips, data translation tools, protein structure prediction
and modelling all combined to make biotechnological R&D far more cost-
effective (Daniell, 1999).

This gradual accumulation of biotechnological insights, instrumentation
and tools was largely driven by the search for new opportunities related to
pharmaceuticals, but their implications are straightforward for food science.
However the two fields form separate research environments not only in
terms of specialized corporate research labs, but also in terms of infrastructure provided by university departments, academic degrees, and government research institutes (GRIs). Lags will occur before food science
absorbs and utilizes advances coming out of pharma-related research.

Our focus on biotechnological R&D in the *food* industry leaves out the
vast research effort directed at genetic modification of crops and animals.
These agroproducts supply the basic raw materials that provide the proteins, fats, carbohydrates and so on, which feed into the value chain of the
food industry. Biotech R&D in the latter is typically concerned with (1)
controlling and modifying the basic ingredients to improve their performance; (2) production of novel ingredients; and (3) processing systems for
the incorporation of ingredients into finished products (Cheetham, 1999;
Jeffcoat, 1999).

We use Lactic Acid Bacteria as a representative case of the impact of
biotech on food R&D because it plays a crucial role in many areas of food
technology. LAB was one of the first organisms used by man to modify
foodstuff (Konings et al., 2000) to achieve preservation, safety and variety
of food, and to inhibit the invasion of other pathogen microorganisms
causing food-borne illnesses or spoilage (Adams, 1999). It is a crucial asset
in modern dairy technologies, including their increasing attention to functional (probiotic) foods. As such, LAB became a natural focus for application and further development of new biotechnological tools as they
gradually became available over the last 20 years (Margolis and Duyk, 1998).
These R&D issues are summarized in Appendix I.

The actual time patterns in food biotech patenting appear in the curves
in Figure 8.2 where a dotted line plots the 180 LAB patents by year of application. The full line, plotted against the right vertical axis, shows all 3425
food-related biotech patents.

During the 1980s few food biotech patents appear (less than 100 per
year), and LAB patenting is sporadic. A move to a moderately elevated
plateau begins around 1990. Towards the end of the 1990s an increase

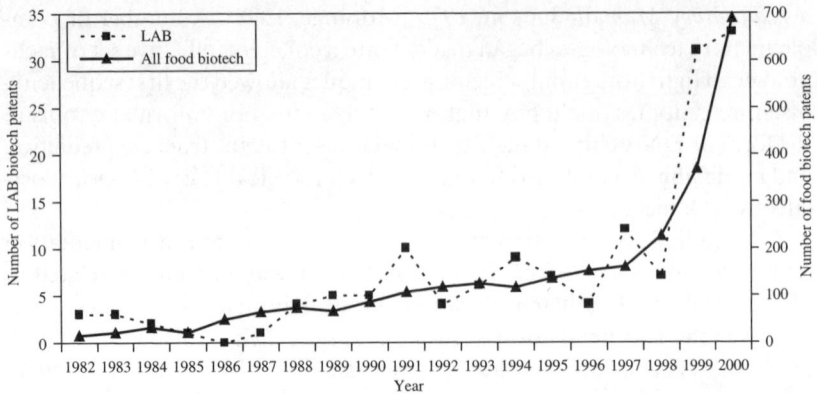

*Figure 8.2   Patents in food biotech and in the subgroup specifically
referring to Lactic Acid Bacteria*

occurs.[3] The early 1990s appear to have brought notable changes to LAB R&D, and to food biotech generally.

The two indicators presented in Table 8.1 indicate a shift in R&D in LAB biotechnology before and after 1992. The first indicator uses the main International Patent Classification (IPC) of each patent to distinguish patents emphasizing novelty in application from patents more oriented towards biotechnology novelties in tools and enablers, resembling the distinction between 'co-specialised vs. transversal research technologies' suggested by Orsenigo et al. (2001). The ratio of tools-oriented IPCs over application-oriented IPCs shows a steep rise from 1992 onwards, reflecting an added emphasis on the integration of an expanding set of techniques and approaches into LAB-related R&D. This interpretation corresponds well with the growing number of inventors from 1992 onwards, observed in the second indicator, reflecting an increase in the specialized skills required to master the diversity of techniques and approaches that become available through the 1990s.

5.2   A Map of R&D Themes

Turning now to a closer look at the cognitive dimensions of LAB biotech R&D, we draw on data extracted from text sections – titles and abstracts – on each patent front page, identifying the novelty claimed and the key principles of how it is brought about. The standard conversion into quantitative representation of this information starts out by selecting keywords. Frequent co-occurrences within patents of keywords signify some common

Table 8.1   Indicators of shifts in LAB-related biotech R&D before and after 1992

Indicators		Periods		All patents
Analytical dimensions	**Definitions**	**1980–91**	**1992–2000**	
Primary orientation of patents towards tools of genetic engineering vs. application*	Ratio of method orientation over application orientation	0.7	4.3	2.8
Complexity of research skills	Av. no. of inventors per patent**	2.3	4.1	3.8

*   'Orientation towards tools of genetic engineering' combines patents that are oriented towards enabling technologies with those introducing novelties or modifications at more generic levels. The classification of patents, based on their International Patent Classification, into these groups is presented in Valentin and Jensen (2002).
** ANOVA test: $p < 5\%$.

dimension of meaning, and patents may be characterized quantitatively on the basis of their affiliation with – or distance from – various dimensions. Data mining tools are now becoming available that allow large numbers of co-words to be considered, enhancing effectiveness in testing outcomes of different ways of bundling multiple co-words into higher order dimensions. This chapter uses BibTechMon software in a procedure accounted for in Appendix III, which also explains the principles behind the key word map presented in Figure 8.3.

To characterize recurrent themes in the 180 LAB patents we use the 973 keywords that are represented by single circles in the map. Proximity between keywords reflects intensity of co-occurrences. Lines represent the strongest 5 per cent of all 46 410 co-occurrences that were used in the iterations to identify 23 themes (statistics on which also are presented in Appendix III).

Not all of the 23 themes lend themselves to meaningful interpretation, and a few of the meaningful themes appear quite randomly across the 180 patents, as indeed they should on the basis of their content. The 12 themes reported on below are those that offer both meaningful interpretation and some level of discriminatory effect between interesting categories of patents.

The 12 themes are summarized in Table 8.2 and they fall into three main groups. Two groups refer to broad areas of application, respectively (i) food process and quality and (ii) pharmaceutical and probiotic functions. The

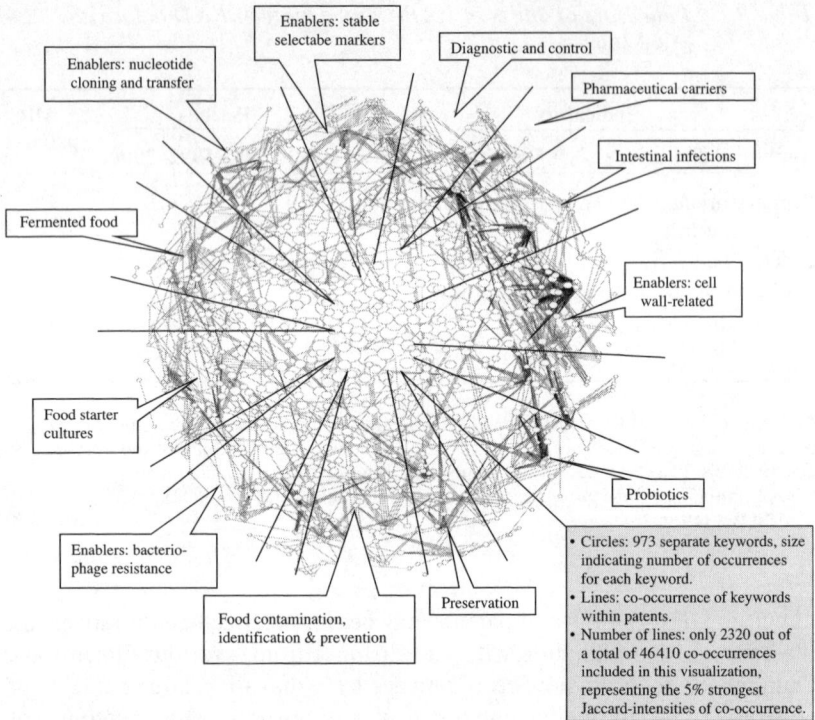

Figure 8.3   R&D themes in 180 patents of LAB-related biotechnology

third group – for want of a better term – is referred to as 'enablers'. Without having the status of generic tools, enablers are techniques that allow effects to be achieved and controlled across multiple applications of LAB-related genetic engineering.

Figure 8.3 positions these themes in a visualization of the keyword map. Each theme is seen to appear as a 'slice' of the map where co-occurrences of its constitutive keywords are particularly dense, indicated by the lines delimiting each theme. Themes in food process and quality (starter cultures and fermented foods) are positioned on one side of the map, opposite the group of pharmaceutical and probiotic functions (intestinal infections, probiotics). The four enablers are located apart from each other, in proximity to the areas of applications to which they are particularly pertinent.

Lines delimiting each theme are merely indicative. Even though themes are concentrated in one particular area of the map, their exact scope may vary. For obvious reasons we would expect areas defined by a field of application

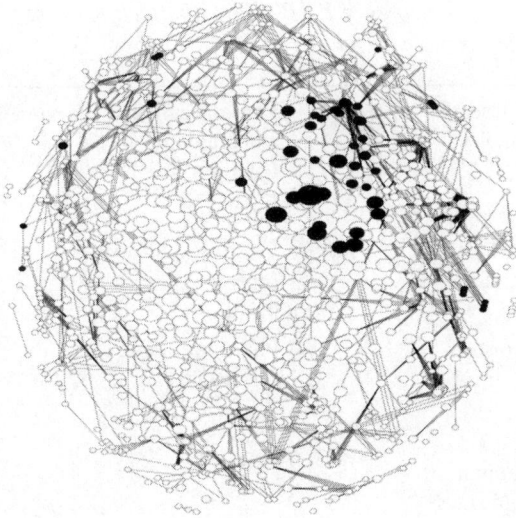

Figure 8.4   Keywords referring to R&D on intestinal infections

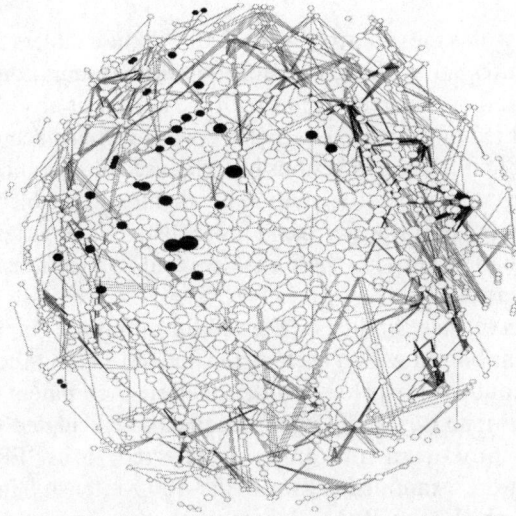

Figure 8.5   Keywords referring to R&D theme on enablers: nucleotide cloning and transfer

Table 8.2   *Selected themes in LAB-related biotech R&D as reflected in*
                titles and abstracts of 180 patents

Theme id	Food process and quality	Theme id	Pharmaceutical and probiotic functions	Theme id	Enabling genetic technologies
14	Starter cultures	1	Probiotics	4	Cell wall-related
15	Fermented foods	6	Intestinal infections	16	Bacteriophage resistance
17	Diagnostics and control	20	Pharmaceutical carriers	18	Nucleotide cloning and transfer
19	Food contamination (identification and prevention)			21	Stable, selectable markers
2	Preservation				

to materialize with a more narrow focus than will the enablers. Exemplifying this point, Figures 8.4 and 8.5 give a more detailed presentation of one application theme and one enabler theme, with the keywords absorbed into each theme highlighted in the map. The theme covering pharmaceutical applications referring to intestinal infections has a focused appearance in the upper 'north eastern' corner of the map. The theme of enablers relating to nucleotide cloning and transfer has a more distributed pattern, indicating particular affiliation between these enablers and applications in fermented foods and in starter cultures.

On the basis of these findings we tested for differences in the configuration of skills mobilized within each of the above 12 R&D themes. Using a procedure accounted for below we identified in each patent the scientists coming from Public Research Organizations and calculated their share of the entire inventor team into an index referred to as 'PRO-intensity'. Correlations were examined between the PRO-intensity index and the intensity with which each of the 12 research themes were present in patents (based on occurrence of their lead keywords). Clearly significant correlations were found, positive with some themes, negative with others. There are strong indications, in other words, of variability in the configuration of skills mobilized in different patents, some themes requiring stronger involvement of public research, other themes requiring less.

The keyword map offers a useful visualization of the decomposability of problem definition. R&D themes in food process and quality are not on one side of the map with enablers positioned on the opposite side. On the contrary, enablers are positioned between areas of application, indicating their interdependencies in the definition of interesting targets for biotech R&D. A closer look at lead keywords (not listed in this chapter) *within* each theme also invariably reveals an intricate mix of process and product attributes.

Furthermore, the map brings out the important observation that pharma-related issues, along with their own set of enablers, are positioned between 1 and 4 o'clock, i.e. independently from the R&D issues in food application of LAB biotech, suggesting both higher coherence *within* these themes and higher decomposability *from* those targeted at traditional food applications. Therefore they also may give rise to differences in distributed innovation. Actors excluded from active problem definition in the nondecomposable innovation space of food, need not also be excluded from the innovation space of pharma-related issues.

To sum up the profile of R&D issues derived form the keyword map:

1. A comprehensive view of all 180 patents during 1980–2000 shows bio-technology being applied to LAB in a number of R&D themes.
2. R&D themes range from processing issues (e.g. starter cultures) to food functionalities (e.g. preservation) and further on to pharmaceutical and probiotic effects. Innovations also include a set of enabling tech-nologies that augment analysis and problem solving across multiple areas of LAB-related applications. R&D issues, in other words, involve not only development of new applications, but also problem solving at more generic levels.
3. Enablers and specific applications are innovated in interrelated forms in positions close to each other in the keyword map, and the keyword mix in each R&D theme reflects an integration of process and product issues. Low decomposability between these themes appears as a key attribute of problem definition in this field of food biotech.
4. However, pharma-related issues, along with their own set of enablers, are positioned opposite from – i.e. independently from – the R&D issues in food application of LAB biotech, indicating the possibility for scientists outside the food industry to play a stronger role in problem definition in this area of LAB biotechnology.
5. Themes differ in the configuration of skills required to carry out innovations, specifically in the involvement of researchers from public science.

These findings offer a richer and more detailed appreciation of a research agenda compared to what we normally obtain from co-word analysis. It tells us why single R&D projects would often require a multiplicity of skills, ranging from process experience, insights into the molecular biology of raw materials, abilities to develop new research tools and concepts, etc. And it informs us of the need for flexible reconfiguration of these skills and experiences from project to project. It also indicates why actors outside the food industry may offer critical contributions to problem solution while at the same time being unable to define relevant innovation problems and targets.

5.3   Evolution of R&D Themes through the 1990s

So far co-occurrences have been considered only from the perspective of how they combine into themes. The next step is to examine how themes map on to the 180 patents from which they have been generated. Statistics on that mapping are presented in Appendix III, Table A8.1, columns 5–6–7. We use this statistic to identify patents that with particular *intensity* express a specific theme, referred to below as 'theme carriers'. In a biological metaphor, these patents equate phenotypic representations of an underlying theme. In operational terms we simply identify patents in the upper median of keyword hits within each theme. With this approach theme carriers within one theme would in some cases inevitably also be theme carriers in others as well. But the procedure is designed to limit such overlaps.

Identifying the year of application for a set of theme carriers informs us of the particular time profile of their theme. We normalize this pattern with an expected appearance (all patents appearing that year as share of total number of patents). Figures 8.6–8.8 give the ratio of observed over expected appearance of theme carriers for each year. Only the time frame of 1990–2000 is considered, since Figure 8.2 revealed merely sporadic activity in the 1980s. The 12 themes considered in this chapter each form a group of main carrier patents. This main carrier group on the average appears also in two to three other themes, at a level corresponding to roughly 10 per cent of their appearance within their main theme.

Figure 8.6 brings together themes in which LAB is used to enhance food processing or food quality. The issues of food fermentation, starter cultures and control of contamination are the 'classic' objectives associated with LAB, and have been so for several millennia (see Appendix I). They reoccur through the 1990s because biotechnology allows them to be addressed and enhanced in novel ways. All four themes throughout the 1990s exhibit moderate fluctuations around expected occurrences (normalized to the value of 1). They follow, in other words, the average rise of patenting activity observed

Figure 8.6 Food process and quality

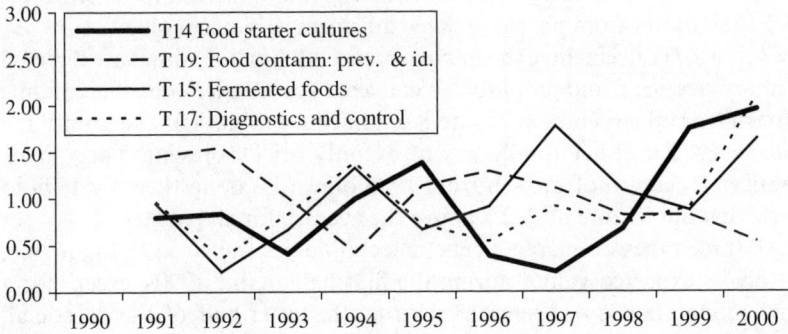

Figure 8.7 Pharmaceutical and probiotic functions

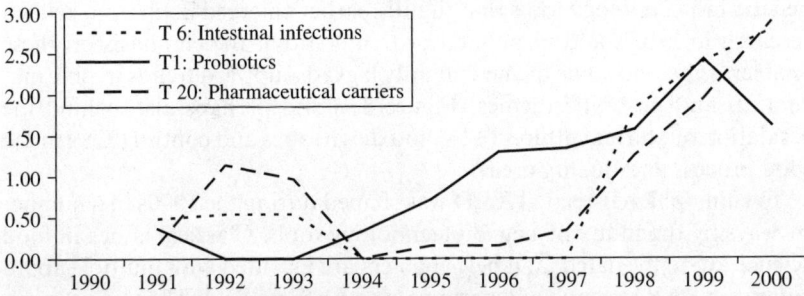

Figure 8.8 Enabling genetic technologies

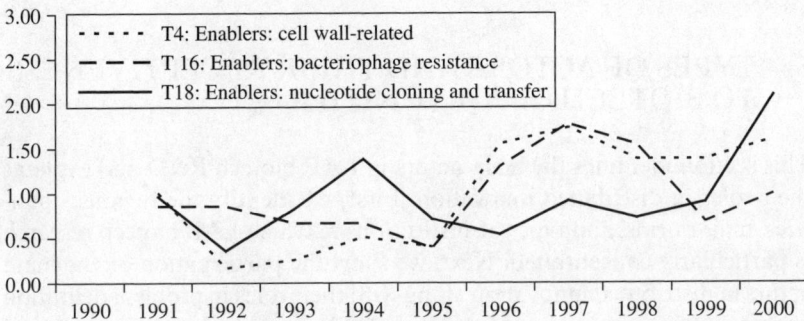

Figures 8.6–8.8 Timing of patenting within three sets of themes in LAB-related R&D 1999–2000. Ratio of observed over expected occurences of top median patents in each year within each theme, two year moving averages

for all LAB issues, as reflected in Figure 8.2. The short upward fluctuations are most likely induced by short waves of exploitation of more generic tools as they spill over from pharma-related advances in biotechnology.

Figure 8.7 collects three themes in which LAB functionalities are directed at pharmaceutical and probiotic objectives. They have a quite different time pattern from the themes in Figure 8.6. Pharma-probiotic themes, until the mid 1990s, are either totally absent or only briefly present. They share, however, a pattern of growth from 1995 onward, taking them by 1999 to levels that are factors of 2–2.5 above the average for all patents.

All three types of enabling genetic technologies, collected in Figure 8.8, are below expected values during the first half of the 1990s, except for a brief moderate rise of theme 18. During the latter half of the decade all three of them at various points grow to levels considerably above expected values. We submit: (1) that the growing presence of these three enablers in LAB-related R&D reflects an inflow into food science of the set of generic biotechnology tools that slightly earlier emerged in pharma-related research in DBFs and in public research; (2) that the emphasis on these enablers is behind some of the – slightly lagged – upward trends in pharmaceutical and probiotic themes (Figure 8.7) and perhaps also behind the escalation of starter cultures (T14) and diagnostics and control (T17) in the food process and quality area.

To sum up: LAB-related R&D was shaped during the 1990s in a number of ways by the influx of new biotechnology tools. Classical issues in food science are rejuvenated, and new enablers are invented with multiple applications in LAB research and perhaps even further afield. The widest implications in the long run perhaps may come from the emerging R&D agenda bridging food, pharmaceuticals and neutraceuticals.

6.   TYPES OF ACTORS AND THEIR RECEPTIVENESS TO BIOTECH OPPORTUNITIES

This section identifies the main actors in LAB biotech R&D and explains their roles in distributed innovation. First, we identify the regions, countries, major firms, and types of institutions in which LAB biotech research is particularly concentrated. Next, we show the participation of the main actors in distributed innovation along with their roles in problem definition and problem solving as revealed in statistical measures derived from patent information. Additional information from documentary sources and case interviews is presented to substantiate these statistical findings. We examine how the roles of the main actors have evolved since 1980, including changes associated with the emergent pharma-related R&D themes.

6.1 The Global Distribution of LAB Biotech Innovations

Tables 8.3 and 8.4 present the geographic distribution of patents in food biotech and in LAB specifically. The lead role of the USA in pharma-related biotech (Allansdottir et al., 2002) is less pronounced in food, where almost 60 per cent of patents are assigned to non-USA organizations. And in the specific field of LAB, roles are completely reversed. The US has merely 9 per cent of patents. Japan has 14 per cent, but Europe clearly has a dominant position with 71 per cent of all patents.

The dominant position of Europe in LAB biotechnology is seen in Table 8.4 to be concentrated in a few countries. The five most active European countries have a global share of LAB patents of 65 per cent, equivalent to 91 per cent of all LAB patents coming out of Europe.

This geographical distribution of LAB patenting not only deviates markedly from the overall pattern of US dominance in biotech, it also brings out the important role played in this field by small European countries. Some specific factors in the five countries help explain their strong presence in LAB biotech.

The leading European countries are those with strong performance in biology and food science (Salter et al., 2000), and quite active science and

Tables 8.3 and 8.4   Geographic distribution of patents in food biotechnology 1980–2000, and the segment thereof relating to LAB

Table 8.3   Global distribution

Global regions	All food-biotech %	All LAB %
Europe	27	71
Japan	21	14
USA	41	9
Others	11	6
Total	100	100
N	3425	180

Table 8.4   European countries

European countries	All food-biotech %	All LAB %	N
NL	13	24	31
UK	23	13	17
FR	16	20	25
CH	7	22	28
DK	11	13	16
Others	30	9	11
Total	100	100	–
N	925	–	128

Note: Nationality of patents refers to the organization(s) to which the patent is assigned. Some 28 patents had multiple assignee organizations coming from the same country. The three patents with assignees from different countries were categorized on the basis of predominant nationality of the inventor team.

technology policies have supported this position. For decades France and the Netherlands have contributed significantly to advances in molecular biology, including specific programmes in sequencing the lactobacillus genome. (Morange, 1998; Roseboom and Rutten, 1998; van der Meulen and Rip, 1998). Denmark has operated a series of science and technology programmes in biotechnology, including a specific subprogram in LAB research (Valentin, 2000).

More importantly however, all five countries are also home to multinational corporations (MNCs) with significant positions in foods and ingredients. Danone and Rodia are present in France, and the Chr. Hansen Group in Denmark. Unilever has a dual base in the UK and in the Netherlands, where we also find companies like Quest International (a subsidiary of Unilever until it was acquired by ICI in 1996) and DSM. The Swiss performance in this field is synonymous with Nestlé patenting achieved by its headquarter laboratories (Boutellier et al., 1999). The location and origin of these MNCs reflects the importance of food products and technologies in their countries. This relationship is examined more thoroughly in Valentin and Jensen (2002) and in Valentin and Jensen (2004).

6.2   Revealed Roles in Distributed Innovation

These national concentrations of LAB biotech innovations and their relations to strong MNCs are useful points of departures for analysis of distributed innovation. This subsection identifies the main actors and the roles they take in problem definition and problem solving as revealed in the ways they are credited in patent documents.

Front pages of the 180 PAB biotech patents list not only parties to whom the patent is assigned, but also the inventors behind the novelty claimed in the patent. A total of 676 inventor references are found in 180 patents. Using various bibliometric sources we identified the host organizations for 625 different inventor participations (51 inventors, mainly from Japan, remain unidentified). A total of 118 *inventor host organizations* were identified. Multiple inventors from the same organization contributing to the same patent may be counted as a single participation, constituting what we refer to as one case of *organizational participation* (OP). Three hundred and twenty such cases of OP were identified. This data allows us to map virtually all participants in the distributed innovation of LAB biotechnology (to the extent they have been patented), in principle permitting us to go down to the level of individual inventor-scientists, but in the present chapter we stay at the level of OPs. The key characteristics of distributed innovation in LAB biotechnology appear in the top seven rows of Table 8.5.

More than two-thirds of the 118 inventor host organizations are PROs (a+f) (letters in parentheses identify rows in Table 8.5). Universities are particularly prevalent, in fact constituting almost half of all inventor organizations. In addition to the three major companies – Unilever, Nestlé and Chr. Hansen – 26 other food/ingredients firms are active. Most of them are in food processing (dairy products primarily) while a few are producers of ingredients. None of them holds more than four patents.

Only 13 OPs and six assignments are recorded for nine different DBFs, i.e. firms specialized in biotech research without downstream production and marketing activities. Compared to the US version of the pharma-related biotech discovery chain (Cockburn et al., 1999) DBFs play a negligible role in LAB biotechnology.

This distribution of inventor host organizations differs starkly from the distribution of assignees (c+g). As the single strongest appropriator, Unilever takes about one-fifth of all assignments. The three largest companies claim 38 per cent, with a similar share going to all other companies. At the other end of the spectrum PROs have a smaller combined share, which breaks down into notable differences between the 6 per cent assigned to universities while the share of GRIs is three times higher. There is roughly a 3–3–2 ratio of assignment shares going to (1) the three large firms combined, (2) all other firms and (3) PROs.

Status in the patent as assignee or inventor host organization also indicates the role of an organization in collaborative R&D. Specifically, we may distinguish between roles in problem processing pertaining respectively to their *definition* and their *solution*:

1. In the majority of the 180 LAB patents, assignee organizations have their own scientists involved in the innovation, but other organizations also contribute to R&D. Typically the patent is assigned to just one of these R&D partners. In an arrangement of that kind we interpret assignments as reflecting a leading role in *orchestrating* the R&D producing the patent and in identifying the problems and the benefits that motivate the R&D project. These assignees, in other words, may be assumed to play a key role in the *problem definition* that gives rise to the innovation project.
2. Organizations contributing inventors, without being assigned the patent, supply the talent, skills and specialized experience required for *problem solving* in R&D leading to patentable innovations.

Even though idiosyncratic circumstances may modify the case for specific patents, this distinction allows us to study patterns of how the roles of problem *definers* (assignees) and problem *solvers* (inventors) are distributed

Table 8.5   *Roles in LAB biotech patenting by type of organization and by specific company, indicators of strength in problem definition and problem solving*

Aspect	Unilever	Nestlé	Chr. Hansen	Other firms	DBFs	GRIs	Universities	N
Patterns of participation								
(a) Inventor host orgs	1	1	1	26	9	25	55	118
(b) Assignee organizations	1[1]	1	1	41	5	18	10	77
(c) Patent assignments[2]	38	26	11	73	6	34	12	200
(d) Organizational participations (OPs)[3]	20	26	9	54	13	79	119	320
(e) OPs including assignee status	19	25	9	53	4	16	9	135
Percentage shares of participation								
(f) Share of inventor host organizations %	0.8	0.8	0.8	22.0	7.6	21.2	46.6	100.0
(g) Share of patent assignment %	19	13	5.5	36.5	3	17	6	100.0
Ratio indicators of strength in problem definition								
(h) Assigned patents per inventor host org (c/a)	38.0	26.0	11.0	2.8		1.4	0.2	1.7
(i) Patent assignments per OP (c/d)	1.9	1.0	1.2	1.4		0.4	0.1	0.6
Ratio indicators of strength in problem solution								
(j) Inventor host org. per assignee org. (a/b)	1.1		1.0	0.6		1.4	5.5	1.5
(k) OPs per participation as assignee (d/e)	1.0	1.0	1.0	1.0		4.9	13.2	2.4
(l) Participations as assignee per patent assignment (e/c)	0.5	1.0	0.8	0.7		0.5	0.8	0.7

Notes:
1 Four different subsidiaries and branches of Unilever are calculated as one unit.
2 Excluding ten individual patent assignments; including multiple organizations sharing same patent.
3 i.e. hosting at least one of the inventors listed in the patent. Multiple scientists from same organization are counted as 1 OP only in each patent.
Excluding 51 inventors whose host organization remains unidentified.

and configured in the 180 patents. These roles are brought out when the five variables in rows a–e in Table 8.3 are calculated into various ratios (presented in rows h–l). Due to the small numbers of DBFs they are omitted from these ratios.

As indicators referring to problem *definition* we calculate average numbers of assigned patents per inventor host organization (h) and per organizational participation (i). All three large firms have high scores on both ratios, indicating their advantages in defining and initiating high volumes of innovation projects. Unilever's score of 1.9 (i) indicates an *additional* strength in initiating – or spotting – commercially relevant research in which they do not directly participate. Universities in particular have low scores on both indicators. Other firms and GRIs perform at a medium level, with each host organization on the average having 2.8 and 1.4 assignments respectively.

On indicators referring to problem *solving* universities top the list. University organizations supply problem solving to patents 5.5 times more frequently than they receive assignments (j), and the level of their participations is 13.2 times higher than their number of assignments (k). The latter ratio is 4.9 for GRI, indicating their proclivity for contributing to problem solving clearly above their level of own assignments.

Since the j-ratio for the three large firms by definition is 1 it is left out of Table 8.3. For 'other firms' this ratio drops to 0.6 reflecting the fact that these assignee firms in many cases have no in-house scientists actively participating in inventor teams, indicating their weakness in problem solving compared to any of the other actors. The three large firms tend to have assignments associated with each of their participations (ratios around 1 in ratio k), but differences in their self reliance as problem solvers are brought out by the last ratio. Only Nestlé has own inventor participations in virtually all patents in which it is an assignee. For Chr. Hansen the ratio is 0.8 while Unilever only has own inventors participating in half of its assigned patents, giving them a position of only medium strength in this respect when compared to Nestlé.

To summarize:

1. LAB biotech R&D requires heterogeneity and recently developed skills beyond what most single inventor organizations can handle internally, rendering distributed innovation the predominant organizational mode for this R&D.
2. Unlike the US style of pharma-related innovations, DBFs are only marginally present in LAB biotech, and the profile of their limited involvement emphasizes contributions to problem solving above *problem definition*.

3. Instead universities are the most preferred type of external partner, and their contribution is focused almost exclusively on contributing solutions to R&D problems that are defined, orchestrated and appropriated by other organizations. In problem solving the role of universities is essential, while in problem definition it is negligible,
4. In this respect GRIs are different. Their overall participation in problem solving is substantial, though not quite as prevalent as that of universities, and it reflects a more balanced potential also for problem definition.
5. All three large firms reveal strength in problem definition. In problem solving Nestlé stands out as the most self reliant organization, while Unilever and Chr. Hansen in this respect are at a medium level.
6. The group of other firms are weak in terms of problem solving, but have medium strength in problem definition.

Table 8.6 recapitulates the strength of each actor as revealed by its share of activities and scoring on the five ratios.

6.3   R&D Profile of Main Actors in LAB Biotechnology

The revealed roles in distributed innovation uncovered above are interpreted in this subsection on the basis of additional information on each of the main actors. This information comes out of documentary sources and in some cases out of interviews conducted with researchers in industry and in corporate labs.

The vertical structure of the food industry gives rise to a particular distribution of R&D across its subsectors and also across different institutions in public science. Each of the subsectors in food processing uses as inputs not only raw materials that are specific for its final products. It also sources

Table 8.6   Roles in problem processing of key firms and main types of actors

Significant firms and types of actors	Dimensions of problem processing	
	Definition	Solution
Chr. Hansen	Strong	Medium
Nestlé	Strong	Strong
Unilever	Strong	Medium
The average firm in food processing	Medium	Weak
Government research institution	Medium	Strong
Universities	Weak	Strong

a complex mix of ingredients that are essential in process regulation and in modifying tastes, structures and other product functions. *Producers of ingredients* deliver these inputs based on quite intensive R&D into process and product technology issues across a broad scope of downstream food products. On this basis, the ingredients sector has come to play a growing role in advancing the knowledge frontier in food technologies (Cheetham, 1999; Jeffcoat, 1999).

The Chr. Hansen Group in Denmark is a niche multinational company, specializing in ingredients for producers of milk-based products all over the world, and is a world leader in cheese ingredients. For more than 100 years LAB has been a crucial microorganism in Chr. Hansen's ingredients and services. To maintain that position, from the mid 1990s the company successfully pursued biotechnological opportunities for further refinement of their ingredients, and today they rank third among companies in the world in terms of numbers of LAB patents based on biotechnology. Chr. Hansen's R&D department is a plentiful point of confluence of information and opportunities, much of which originates from the clients' process problems (Valentin, 2000). However, this information translates into interesting innovation targets only when brought together with Chr. Hansen's own biological understanding of possibilities for modifying LAB functionalities, giving problem definition a nondecomposable quality.

A useful example of what low decomposability of problem definition means in this context came out of the case studies we undertook to understand the research behind LAB patents. Based on its long experience with supplying ingredients to cheese manufacturers, Chr. Hansen is aware not only of the economies to be gained from reduction in cheese maturation time. They also know that the process will benefit from and be susceptible to acceleration only at certain stages. They have a deep understanding of the maturation process as a degeneration of milk proteins handled by a set of enzymes, the numbers and functions of which could be controlled by promoters. This confluence of experience and insight allowed Chr. Hansen to identify effective management of precisely these promoters as a highly relevant target for biotech research, and it led to a problem definition that could not have evolved from separate deliberations of its constituent components of knowledge and information. Once the problem was properly defined, however, Chr. Hansen pursued swift problem solving though distributed innovation involving not only their own researchers, but also the expertise of several public research partners. Collaboratively they developed (1) a method for identification of the promoters and (2) enabling tools by which the function of the promoters may be controlled (source: own interviews).

In this case *problem solving* obviously had a level of decomposability allowing it to be successfully pursued in a collaborative research project. *Problem definition*, however, was the result of a nondecomposable process.

This pattern in Chr. Hansen's processing of innovation problems gives it a strong position in problem definition. While it undertakes more R&D than the average food company, it still has a considerably smaller volume compared to Nestlé, which accounts for its medium level strength in biotech-based problem solution observed in Table 8.5.

As very large MNCs Nestlé and Unilever have sizeable internal R&D resources at their disposal. This allowed them to enter early into biotech applications within their product lines, and they also have the sophistication and volume of R&D to undertake large scale external research collaboration. However, their exact specialization differs in ways that also translate into dissimilar R&D agendas in biotech. For more than a century Nestlé has specialized in milk-based products, and over the last decades its competitive profile has increasingly emphasized nutritional qualities, backed by advanced internal R&D (Boutellier et al., 1999). Nestlé has a strong presence in biotech associated with these issues, making it much less dependent on external R&D collaboration. In a previous analysis of this data set (Valentin and Jensen, 2004) we demonstrated that up until the mid 1990s Nestlé carried out their LAB biotech R&D as internal research only. Its shift to distributed innovation seems to be associated with an increasing attention to the emerging agenda for pharma-related applications of LAB biotechnology (Pridmore et al., 2000). In this novel agenda Nestlé's interest in nutritional research appears to offer a new set of advantages, but of a kind that are additionally enhanced by external collaboration.

Unilever in the 1930s arose as a merger of British production of soaps with Dutch activities in margarine. The product line has since diversified further into a variety of frozen and canned foods (ice cream, fish products, precooked meals, etc.), and home and personal care products. Unilever's R&D is correspondingly diverse, organized along major product types (Unilever home page, 2003). Within each of these R&D specializations the emphasis on product and market focus builds strong positions in problem definition. But concentration of R&D on diverse applications makes Unilever more dependent on contributions from external research into a highly heterogeneous array of biotech applications, accounting for its medium level strength in problem solving observed in Table 8.3.

Firms in food processing are traditionally based on specific raw materials (diary products, meat products etc.) and undertake R&D on a limited scale, often narrowly focused on particular parameters of quality, variability or hygiene of raw materials and final products (Senker, 1987). In most cases this R&D profile prevents them from building in-house expertise capable of following and exploiting the advances of biotechnology (Kvistgaard, 1990). As a consequence, in problem solving their position tends to be weak, making them quite dependent on outside expertise in collaborative arrangements. Their deep experience in integrated product process issues offers opportunities for problem definition, but only in areas pertaining to their specialization in products and raw materials. In this respect they are also constrained by their narrow R&D focus, accounting for their revealed level of medium strength in problem definition.

Government Research Institutes (GRIs) and universities represent two quite distinct profiles in LAB food biotech with implications for their positions in problem definition and solution. Prior to World War I most countries established GRIs that specialized in food safety and quality. Due to its implications for public health, in particular tuberculosis, its handling and processing of milk in agriculture and dairies also ranked high on the agenda of these GRIs. Furthermore, their science was needed to back the formulation of standards and regulations to handle complex interdependencies in the value chain of milk comprising farms, transport, processing, distribution and consumption (Rosenberg, 1985). This mandate required then – and still does today – ongoing research into industrial process product interdependencies to an extent found in very few other areas of public science (Leisner, 2002). This gives GRIs a strong role in LAB-related problem solution, but also some standing in problem definition, reflected in their position at a revealed medium level in the latter.

Universities are the major source of researchers capable of translating recent advances in global molecular biology into problem solving skills and experience. This makes them highly useful collaboration partners in problem solving in biotech innovation. Their remoteness from food product and process issues creates obvious disadvantages when it to comes to problem definition. However, in areas where LAB biotech research diversifies into issues where information on opportunities and targets flow in decomposed forms in the public domain, universities could come to play an increasing role also in problem definition. That is precisely what characterizes the issues now emerging in pharma-related applications of LAB biotech (cf. Figure 8.7). The decomposability of problem definition associated with these new issues gives university research possibilities for a more aggressive role in problem definition, quite different from the weak position in problem definition until now (as revealed in Table 8.5).

To conclude, information from documentary sources and case interviews from each of the main actors produce profiles of their R&D that are consistent with their revealed roles as problem definers and problem solvers in LAB biotech R&D as summarized in Table 8.6.

6.4 Timing

From the above presentation of findings it cannot be ruled out that the main actors might have *shifted their roles* in distributed innovation over the two decades covered by our time frame in ways that could affect the main hypothesis of this chapter. Could it be, for instance, that GRIs, universities or DBFs in the initial breakthrough phase had a higher level of significance, which disappears when data from early years are collapsed with data on the much higher volume of activity in later stages?

Higher significance for these organizations in the early phases would bring this field of innovations closer into conformity with standard arguments from the technology cycle literature that the weight of activity shifts from small entrepreneurial units to larger firms as the cycle unfolds (Tushman et al., 1997; Utterback, 1994). It would also make the LAB biotech case more similar to observations on pharma-related biotechnology where successive waves of incoming new DBFs have restructured and redistributed tasks in discovery-oriented R&D (Orsenigo et al., 2001). It would weaken the main argument of this chapter, that different levels of decomposability of problem definition and problem solving have assigned roles in distributed innovation for all main actors since biotechnology entered this field of R&D in the early 1980s.

To examine the patterns across time, Figure 8.10 plots patent applications by years for a categorization of actors similar to the one used in Table 8.5. To bring out underlying trends more clearly, five year moving averages are applied.

Table 8.5 showed an overall 3–3–2 proportion of assignments for the following three groups: (1) the three top patenting companies, (2) other companies and (3) PROs. Figure 8.9 shows that these proportions by and large prevail over the two decades, but it also uncovers some differences in timing of activities for the three large companies: Unilever begins an increase in patenting activity in the late 1980s. Nestlé begins to increase in the early 1990s with activities still expanding in the late 1990s (in fact at that time outgrowing Unilever's level of patenting). Chr. Hansen does not become active until the mid 1990s.

A different angle on the growth of patent producing R&D is presented in Figure 8.10, which distinguishes *first participation* for any of the 118 inventor host organizations from contributions from organizations with reoccurring participation in LAB patent inventor teams.

Throughout the 1990s the increasing volume of LAB patents is based primarily on reoccurring participations. Each year 15–20 organizations with previous LAB patenting experience reoccur as co-inventors in new patents. About ten organizations have their first participation.

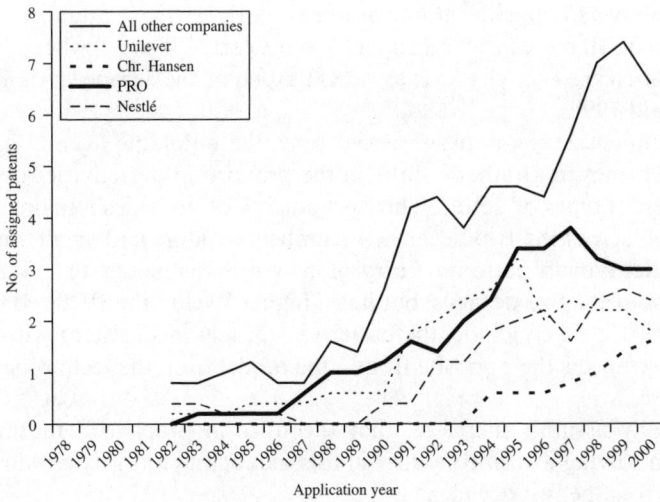

Figure 8.9   Application year for patents for different types of organizations, five year moving averages

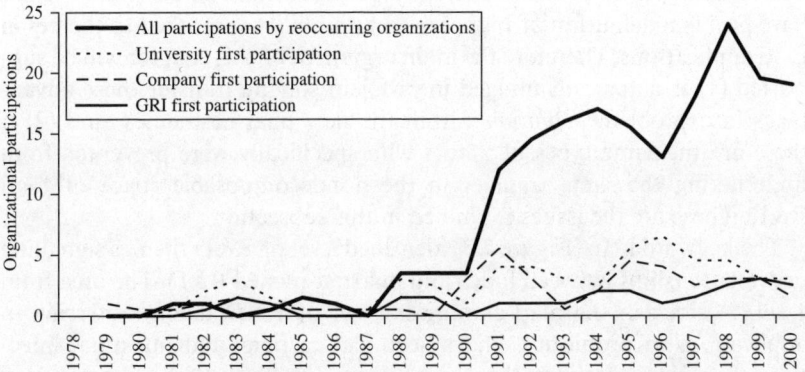

Figure 8.10   Organizational participation by type of organization and by first vs. reoccurring participation, two year moving averages

The breakdown by organizational types shows for the 1990s an inflow of two to four new companies every year, rising slowly through the decade. From the previous section we know that firms invariably enter as assignees. And from previous examination of the data (Valentin and Jensen, 2004) we know that they tend to bring their 'own' university partners with them in a collaborative arrangement, explaining most of the elevated level of university entries observed through the 1990s. During certain intervals the inflow of new university participations becomes particularly pronounced, e.g. (1) when the LAB biotech agenda opened in the early 1980s; (2) when it shifted into an increased activity level towards the end of the decade, and (3) again in the mid 1990s.

To summarize the patterns across time, the unfolding agenda of LAB biotech brings no dramatic shifts in the proportion of activities observed for different types of actors. Their proportion of activities remains largely the same across the two decades. Incumbents – large and small – expand in parallel growth patterns. Entry of new firms is moderate, takes place throughout the two decades, but has a higher level in the 1990s compared to the 1980s. To conclude, the pattern is entirely inconsistent with – and in some respects the opposite from – the model from the technology cycle literature.

We may assume, therefore, that the roles in problem definition and problem solving identified earlier in this section have largely remained the same across the two decades.

6.5 An Emerging Fusion of Food and Pharma R&D

For reasons presented below the emergent pharma-related R&D themes have problem definition of higher decomposability compared to themes in food applications. Therefore the main argument of this chapter will be supported (1) if actors advantaged in problem solving transfer these advantages into problem *definition* within the new pharma-themes, and (2) if these are the same types of actors who specifically were prevented from undertaking the same transfer in the nondecomposable space of food R&D. These are the issues examined in this subsection.

The keyword map (Figure 8.3) identified a set of R&D themes signalling new relationships between food- and pharma-related R&D. The area from 1 to 4 o'clock in the map comprises a set of R&D themes referring to applications in probiotics, pharmaceutical carriers, and intestinal infections. Positioned between these applications we find enabling innovations relating to cell walls and their significance for immune response. These themes represent a crossover from food to pharmaceutical research themes, and they have all been subject to rising attention from 1995 onwards (Figures 8.7 and 8.8).

Pharma-related innovation builds on problem identification of a more decomposable quality compared to foods. In pharmaceutical R&D clinical research accumulates into a highly articulated structure of information and knowledge concerning effects and side effects of drugs. A search architecture in the public knowledge domain (cf. competitor intelligence service products like IDdb3 (the Investigational Drugs Database (IDdb), 2003) gives

pharmaceutical research highly effective access to knowledge on functionalities, although pharmaceutical firms, of course, guard their specific
insights on *new* targets in the pipeline.

If our main argument holds, these differences in problem definition in
pharma and food biotech innovations translate into different patterns of
distributed innovation. Specifically, universities should be better positioned
to define pharma-related innovation problems as reflected in a higher share
of patent assignment. We also would expect them to exploit those opportunities in the areas of their particular advantage, i.e. in innovation of
enablers and not in specific applications.

Furthermore, we would expect food companies to be disadvantaged by
the novelty which pharma-related applications would represent to them.
The one exception would be Nestlé given their clear priorities in nutritional
and health-related food research.

We examine these propositions using standardized keyword scores characterizing the affiliation of each patent with each of the 12 R&D themes
(presented in Appendix III, Table A8.1). First, ANOVA procedures were
applied to test differences between assignee groups in their average keyword
scores on each R&D theme. Table 8.7 shows that the few patents that are
assigned to universities differ starkly from those assigned to all other groups
precisely by being significantly more strongly affiliated with the theme of cell
wall-related enablers. Patents assigned to Nestlé have significantly stronger
affiliation with the themes of probiotics and pharmaceutical carriers.

Second, we test if patents affiliated with the new pharma-related themes
are distinguished by particular compositions of their inventor teams. Shares
of inventors coming from each of the five groups (universities, GRIs,
Nestlé, Unilever, and other companies) were calculated for each patent.
These relative measures were correlated with keyword scores for each of the

*Table 8.7   Differences in keyword scoring on R&D themes between types of
assignee organizations (no. of organizations)*

Theme	Highest scoring organization type	All other organizations
4. Cell wall-related enablers***	Universities: 0.14 [9]	0.03 [163]
1. Probiotics***	Nestlé: 0.10 [26]	0.02 [146]
2. Pharmaceutical carriers***	Nestlé: 0.11 [26]	0.04 [146]

Notes:
ANOVA test: p < 0.01.
Keyword scores for R&D themes are standardized to values 0–1.
Asterisks indicating significance levels from Gabriel's multiple comparison procedure.

*Table 8.8   Revealed theme affiliation of patents correlated with the share
of inventors for different organizations 1995–2000*

R&D themes	Inventor shares in patents	Pearson correlation coefficients
4. Cell wall-related	Universities	0.22**
1. Probiotics	Nestlé	0.31***
20. Pharmaceutical carriers	Nestlé	0.20***
No. of patents		*95*

Notes: ** statistically significant at the 0.05 level, *** statistically significant at the 1.0 level.

pharma-related themes for the period from 1995 onwards, i.e. the point
from which they receive increasing attention in LAB biotech R&D.

The findings reported in Table 8.8 confirm a particular involvement
of universities and of Nestlé in three out of these four pharma-related
R&D themes. No other groups showed any systematic affiliation with these
themes, indicating that Nestlé and universities play a key role in opening the
crossover from food- to pharma-related R&D themes observed from 1995
onwards.

To summarize: the emergence of a pharma-related research agenda in LAB
biotech permits us to examine if the higher decomposability of problem definition in this new field brings modifications to the organization of distributed
innovation observed for food application. Findings confirm that university
scientists exploit the opportunities of a more decomposed space for problem
identification by undertaking their own orchestration of collaborative
research. Findings also confirm that firms and GRIs cannot transfer their
advantage in the nondecomposable problem space of food to the new problem
space of pharma-applications. The notable exception is Nestlé due to their
long research tradition which has prepared them precisely for this crossover.

7.   CONCLUDING REMARKS

This chapter has reported on the organization of distributed innovation
shaped by the major discontinuity in the life sciences and their associated
technologies that has unfolded over the past three decades. While most
studies have focused on its effects on pharmaceutical R&D, this chapter
studies food processing technologies, taking biotech exploitation of
the ubiquitous microorganism of Lactic Acid Bacteria as its example.
Comprehensive recording of all 180 innovations patented in this field allows

us to build a complete map of contributions from research organizations to these innovations, from which we may reconstruct their pattern of collaboration and its evolution. Using textmining methodologies on patent titles and abstracts we identify the major R&D themes and their evolution through the 1990s. A partial fusion of food R&D with nutraceutical and pharma-related issues emerges towards the end of the decade.

Throughout their adjustment to this discontinuity incumbents largely maintain positions and proportional shares of activity. Twenty-six firms in food processing (dairy products in particular) each take out a few patentable innovations based on their own active involvement in R&D. Large incumbents like Unilever and Nestlé patent at a level tenfold higher. More than 100 research organizations – most of them university departments – become involved in the collaborative R&D behind the 180 patents. But the patents to which they supply critical skills and experience are rarely assigned to them. The few cases of university assignments are virtually all in the recently emerging pharma-related R&D agenda. GRIs, however, are much more frequently assigned patents to which they contribute R&D. DBFs play a negligible role.

To explain the organizational characteristics of this distributed innovation we suggest a distinction between *definition* and *solution* of innovation problems. While the latter in food biotech innovations has high Simonean decomposability the former has not, giving incumbents considerable advantages in opportunities for recognizing and assessing the economic prospects of using biotechnology to augment food technologies. University scientists may contribute critical problem solving skills, but are by themselves unable to identify valuable innovation targets. Only the pharma-related R&D themes emerging from the mid 1990s offer to university scientists spaces for problem definition of sufficient decomposability to allow them the role of R&D orchestrators. That precisely becomes the R&D agenda in which they provide the cognitive ordering of collaborative projects that make them assignees of resultant patents. The research mandate of GRIs in this field, on the other hand, allows them to share much of the combinatorial cognitive advantages of food firms, and they orchestrate and appropriate their own R&D accordingly.

7.1   Results

Different types of results – empirical, methodological and theoretical – are generated in this chapter. Empirically we demonstrate how an industry and its incumbents largely maintain structure and positions while a technological transformation unfolds in their underlying knowledge bases. It offers a clear exemplification of Pavitt's reminder to us that technological and industrial/corporate transformations are two very different phenomena (Pavitt, 1998). Precisely the multitechnological nature of firms (Granstrand et al., 1997) allows them to absorb new technologies gradually and in forms permitting them at the same time to capitalize on strengths in other capabilities.

The subtle interrelationships between technological and corporate transformations in no way detract from the importance of understanding discontinuous innovations. But it has the important *methodological implication* that technological change may be observed through the lens of corporate change only if we accept considerable levels of noise and distortions. The study of innovations therefore needs methodologies for mapping of single technologies and their evolution, decoupled from their corporate frameworks. Without offering a complete decoupling, patent data in this regard take us a valuable step forward. In this respect the present chapter suggests methodologies for extending patent data with identification of their inventors and their host organizations and with text mining of their R&D issues.

Theoretically the distinction introduced in the chapter between problem decomposability as referring *separately* to their definition and their solution has implications not only for the analysis above but also more generally for understanding competence enhancement. The literature is not always clear on whether enhancement of firms' competencies in innovation refers to their relation to complementary assets, and hence essentially to their appropriability (Teece, 1986), or refers to the innovation process proper. In this respect the present chapter submits a theoretical argument on decomposability of definition of innovation problems as an attribute of the innovation process proper, not of its forward linkages to complementary assets. This decomposability argument in turn specifies conditions under which firms may remain favourably positioned to extract knowledge and skills – potential value – from a widely distributed network in subsequent problem solving.

7.2 Discussion

Distributed forms of innovation materialize in response to strong underlying forces. Increasing costs and commercial risks of R&D, the demand on firms to master a broadening range of diverse technologies, increasing complexity and multidisciplinarity of technologies, and shortening product life cycles all render distributed innovation increasingly significant as an organizational vehicle for technological and economic progress (Coombs and Metcalfe, 2000).

For that reason it is important to understand what shapes the division of tasks between the key actors of distributed innovation such as producers of goods and services (firms), suppliers of abstract knowledge (universities), translators of knowledge into new fields of applications (DBFs and GRIs) and providers of capital.

The business press tends to see a particular rendering of the US model for distributed innovation as the ideal framework for high tech growth. This model emphasizes market-based formation of small science-based firms (DBFs), backed by venture capital and strong basic science. The inference is quickly drawn that other countries, to get their share of high tech growth, must emulate the US model. The reservation that this model operates for some US high tech sectors, but not for others, is neglected, as is the fact that even *within* technologies new entry firms may be critical for certain types of US high tech growth, while they are immaterial for others (Cockburn et al., 1999).

The findings of the present chapter suggest that the US package of scientist–entrepreneurs and competent venture capital offers powerful comparative advantages only to certain types of high tech activities. For all we know, other types of high tech may thrive better under different conditions. That seems to be the case for the area of food biotech examined in this chapter. The theoretical argument on decomposability suggested here indicates some of the attributes in the institutional framework that most likely would benefit *this* exploitation of the biotech discontinuity.

Low decomposability of problem definition in this field leaves little room for science-based start-ups specialized in innovative research. At the same time, problem solving requires confluence of heterogeneous fields of research, all of which are continuously affected by steep progress in the science frontier they are part of. With these conditions for innovations, commercial food biotechnology will benefit less from policies promoting formation of DBFs, and will thrive on access to the benefits of strong, responsive and multifaceted *public* science. To deliver those benefits public science would need the heterogeneity represented in our data by the different mandates differentiating universities from GRIs; and it would need incentives and institutional differentiation conducive to the balanced role of public scientists of being committed both to scientific progress *and* to responsiveness to technological and commercial challenges.

APPENDICES

Appendix I: Lactic Acid Bacteria in Food Science and Technology

Lactic Acid Bacteria (LAB) was one of the first organisms used by man to modify foodstuff (Konings et al., 2000), achieving preservation, safety and variety of food, and inhibiting invasion of other pathogen microorganisms causing food-borne illnesses or spoilage (Adams, 1999).

To yield cheese, yoghurt and other dairy products LAB ferments milk by decomposing lactose (the main saccharine in milk) to generate a carbon source and to get energy. Rather than breaking down lactose completely LAB leave lactic acids as one of many by-products. Lactic acid reduces pH in milk, leading it to sour and to form the familiar thick texture of butter-milk and yoghurt. Following acidification the process of adding flavour and aroma is started by adding a starter culture composed of a variety of LAB strains. The production of cheese begins with an identical procedure of acidification, but then adds an enzyme treatment of the milk protein casein to generate a creamy lump (curd), which forms the basis for the subsequent processes.

LAB plays a crucial role in modern production of fermented dairy prod-ucts, vegetables and meat, as well as in the processing of wine products. Over the last decade scientific understanding of LAB (e.g. its metabolism and functions) has expanded considerably, opening up the way for more reliable process control in production and for an increasing range of indus-trial applications, including its use in food as additives.

Expanding applications also include explorations of LAB in dairy prod-ucts enhancing probiotic functions (i.e. favourably affecting the microbio-logical flora in the gastrointestinal tract of humans or animals). Probiotic effects of different members of the LAB family, for instance lactobacillus, have been shown to appear not only in intestinal microflora (Berg, 1998; Dugas et al., 1999; Roberfroid, 2000; Saarela et al., 2000; Tannock, 1997) but also in the immune system (Reid, 1999; Wagner et al., 1997). These new fields of application of lactic acid bacteria are promising targets for future research, which will gain further momentum from the growing under-standing of the genomics of the gram-positive bacteria.

Fermentation of yoghurt and hard 'cooked' cheese products like Emmenthal, Parmigiano, Grana etc. requires incubation of milk or curd at temperatures above 45°C, under which conditions normal versions of LAB are unable to survive (Delcour et al., 2000). Specific strains of the bacteria with thermophilic characteristics can survive at this elevated temperature. Compared to other areas of LAB research, thermophilic strains have received less attention, until a steep increase in efforts over the last few years focused on genetics, metabolism and physiology gave notable results in terms of molecular tools and knowledge. Particular attention has been given to research on bacteriophages, stress response and polysaccharides exported to the culture medium (the fermented milk product). Research in thermophilic bacteriophages has special significance because they are key drivers of instability and costs in the dairy industry.

Other recent foci for research have addressed optimization of LAB culture growth and resistance towards 'phages, and the development of sustainable strains from a biological safety perspective (de Vos, 1999; Saarela et al., 2000). Nevertheless we are far from a complete understanding. New insights emerging from research into the genomics of LAB strains are expected to generate new molecular tools for researchers in the field (Kuipers, 2000).

Appendix II: Patent Search Procedures

To identify patents in the intersection of 'food' and 'biotechnology' we searched online databases using combinations of International Patent Classifications (IPCs) and text strings in patent titles and abstracts. Patent searches give the researcher an unavoidable wide margin of choice. One crucial trade-off is that strict definitions bring distinctness to the type of science and technology actually selected, but they achieve far smaller search outputs than do more open ended criteria. Our empirical design includes comparisons between different innovation systems engaged in LAB R&D and between LAB and other food biotech. This comparative design requires priority to consistency above volume, and thus application of restrictive criteria. For each patent meeting our search criteria we identified its initial version and used it throughout the study to represent its entire patent family.

Searches carried out in the autumn of 2001 using these procedures generated a total of 3425 patents in food biotechnology for the period 1976–2000. Less restrictive criteria could have increased that number by factors of 2 or 3, but would have made it more ambiguous in what sense the patents related to food technologies, or in what sense new biotechnology played a role in its development (a similar approach has been used to target specific fields within optoelectronics (Miyazaki, 1994)).

Patents were first identified and selected in Derwent World Patent Index on the basis of IPC and text string criteria. Patent number, all IPC classes, assignee names and inventor names also were recorded from Derwent. esp@cenet was used for identifying nationalities of assignees and inventors, and for recording the distinction between main and secondary IPC classes. A third database, Delphion Intellectual Property Network, was used in a number of cases for achieving information lacking in the two other sources.

In patent families the initial patent was identified, and throughout this study information refers to this initial patent, with the exception of IPCs where we recorded all classifications accumulated from consecutive examinations involving the patent family from e.g. national, European and US examinations.

Data collection was carried out in two steps, beginning with patents in food biotechnology. A total of 3425 patents (families) were identified, involving both genetic modification AND food technologies, as operationalized in the following criteria:

1. *Criteria defining genetic engineering*: Patents should involve genetic engineering defined as modification and introduction of new genes to achieve e.g. immunological effects or to develop new enzymes or formation of new starter cultures with altered metabolic properties. New enzymes are included because in their modern version they are based on extensive use of genetic engineering techniques. In operational terms patents were screened for: (A) IPC numbers C12N15*, C12N1/15, C12N1/19, C12N1/21, C12N5/10, C12N5/12, C12N5/14, C12N5/16, C12N5/18, C12N5/20, C12N5/22 and C12N7/01. Or (B) Texts strings relating to genetic engineering in patent title and abstract.
2. *Criteria defining food*: (A) IPC numbers A23* or A21D* or A22B* or A22C* referring to foods or foodstuffs and their treatment. Or (B) Title or abstract text strings referring to targets for food R&D such as wine, beer, meat, fruit, fish, poultry, fat, eggs, vegetable, butter, cocoa, dough, flour, curd, nutritive, additives, probiotic, cream, cereal, cheese, yoghurt, milk, food.
3. *Criteria on LAB relevance*: Patents should relate specifically to lactic acid bacteria in terms of being concerned with either its properties, with applying them to achieve specified further derived effects, or with using them to establish new tools, approaches or instrumentation in genetic or molecular biology. In operational terms patents should meet the criteria of (A) having the IPC number referring specifically to lactobacillus subtypes/derivatives thereof (C12R1/225, C12R1/23, C12R1/24, C12R1/245 or C12R1/25), or (B) having in their title or abstract text strings referring to LAB or subtypes/derivatives thereof.

Using bibliometric sources and Internet searches we identified the host organization of each inventor identified in each patent. Most patents have inventors from multiple organizations, frequently in configurations of companies, universities or GRIs, thus offering information on significant organizational aspects of its underlying R&D projects.

Appendix III: Applying BibTechMon Software to Generate Research and Technology Themes across 180 LAB Patents

To identify common themes in the research and technology issues addressed in the 180 patents we examine co-occurrences of key terms, using BibTechMon data mining software (for an introduction see e.g. Noll et al., 2000).

Our initial entry of titles and abstracts of 180 patents into the database generates more than 10 000 separate terms, counting one appearance only of each keyword in each patent. All redundant terms (like 'and, or, else, if', etc.) are deleted, as are terms that in our particular sample would be incapable of generating differences in meaning or final interpretation. The latter include terms like 'contain, mol, concentration, solution'. This reduced set of 2095 terms is standardized to handle differences in spelling, abbreviation, synonyms etc., bringing us to a final set of 973 terms, hereafter referred to as keywords.

The network is generated by a co-word analysis, calculating the intensities of all relations between keywords. The intensity of a relationship between any two keywords reflects how frequently they appear together in different patents. This generates a co-occurrence matrix that is normalized using the Jaccard Index given by the equation:

$$J_{ij} = \left[\frac{C_{ij}}{C_{ii} + C_{jj} - C_{ij}} \right]$$

where C_{ij} are co-occurrences of keywords i and j, and C_{ii} is the total number of occurrences of keywords i.

Using the procedure presented in Dachs et al. (2001) the two dimensional map presented in Figure 8.3 is generated, with proximity between keywords reflecting the Jaccard intensity of their relationship.

In the keyword network we identify *themes*, defined as configurations of keywords connected by their highest Jaccard intensities. The formation of each theme takes its point of departure in a core of highly connected keywords. Keywords are added successively on the basis of their Jaccard intensities with the core configuration (applying the Shell facility offered by BibTechMon) down to an intensity level of 0.20 (thus leaving 72 keywords (out of a total of 973) unaffiliated with themes). Each keyword contributes to one theme only. Through this procedure a total of 973 keywords are categorized into 23 different themes.

Table A8.1 (1) shows themes comprising an average of 42 keywords, each of which has an average occurrence of 3.15 (3). The 23 keywords ranking as the most frequently occurring in their respective themes have an average occurrence of 12.6 (4).

Each patent may be characterized by its number of 'hits' among all keywords or among the subset of keywords within single themes. The most 'keyword-intensive' single patent in each theme averages 19 keyword hits.

On average the themes are totally unaffiliated (keywords scoring = 0) with 123 patents (6), i.e. 57 patents having some level of positive scoring within single themes. Column 7 gives patents located in the top median of this distribution, referred to as 'theme carriers'. For all 23 themes, the average size of the main carrier group is 18 patents. The average for the 12 themes included in the analysis in this chapter is 24 patents.

Table A8.1   Statistics on 23 R&D themes

Theme	Keyword (kw) characteristics				Patent characteristics		
	1 Sum of kw per theme	2 Sum of kw occurrences in all patents	3 Average occurrence	4 Occurrence of most frequent kw	5 Max. occurrence of kw in a single patent	6 Patents with 0 kw occurrences	7 Patents in top median of 2
1	37	149	4.03	16	26	116	20
2	31	68	2.19	9	20	149	10
3	55	115	2.09	9	23	138	15
4	56	255	4.55	18	37	92	26
5	62	123	1.98	7	32	136	13
6	58	166	2.86	16	34	115	15
7	39	162	4.15	19	20	106	20
8	42	130	3.10	10	18	118	21
9	33	94	2.85	8	11	130	11
10	53	124	2.34	8	25	129	14
11	53	127	2.40	8	11	125	13
12	42	168	4.00	12	23	112	33
13	40	123	3.08	18	14	121	21
14	52	136	2.62	15	19	119	20
15	33	129	3.91	10	20	122	20
16	40	122	3.05	8	17	124	20
17	42	127	3.02	16	19	123	24
18	37	111	3.00	11	10	127	18
19	51	147	2.88	12	10	117	16
20	42	169	4.02	21	12	111	25
21	20	79	3.95	14	8	132	11
22	31	107	3.45	13	19	136	13
23	24	72	3.00	8	13	139	10
Sum	973	3003	72.53				
Average	42	131	3.15	12.64	19.17	123.35	17.8

NOTES

1. We received useful comments from Keld Laursen. Research for this chapter has been supported by 'Innovation in the Øresund Region', a research program funded by ØFORSK and by 'Centre for Interdisciplinary studies in Management of Technology', funded by the Danish Social Science Research Council.
2. In a more subtle sense, however, decomposability of complex problems is affected by changes in the cognitive attributes of their elements. Take an example where a set of four elements (A, B, C, D) defies decomposition into two simpler subproblems (A,B) and (C,D) because of significant nonspecified interdependency between the two subsets. Improved understanding of this interdependency – e.g. in the form of the diffusion curve in Boisot's I-space (Boisot, 1998) – will specify relationships in forms that will also increasingly allow decomposed problem processing. In this way, decomposability is indirectly contingent on the level of scientific understanding of both its elements and their interdependencies, as indicated also in Simon's original formulations.
3. The time pattern in Figure 8.2 fits well into a more generalized theory of the dynamics of science-driven technologies (Grupp, 1998; Valentin and Jensen, 2002).

REFERENCES

Adams, M. R. (1999), 'Safety of Industrial Lactic Acid Bacteria', *Journal of Biotechnology*, **68** (2/3), 171–8.

Allansdottir, A., A. Bonaccorsi, A. Gambardella, M. Mariani, L. Orsenigo, F. Pammolli and M. Riccaboni (2002), *Innovation and Competitiveness in European Biotechnology*, Brussels: Enterprise Directorate General, European Commission.

Anderson, P. and M. L. Tushman (1991), 'Managing through cycles of technological change', *Research Technology Management*, **34** (3), 26–31.

Arora, A., A. Gambardella and A. Fosfuri (2001), *Markets for Technology and their Implications for Corporate Strategy*, Cambridge, MA: MIT Press.

Berg, R. D. (1998), 'Probiotics, prebiotics or "conbiotics"?', *Trends in Microbiology*, **6** (3), 89–92.

Boisot, M. H. (1998), *Knowledge Asset: Securing Competitive Advantage in the Information Economy*, Oxford: Oxford University Press.

Bonaccorsi, A., F. Pammolli, P. Massimo and S. Tani (2001) 'Nature of innovation and technology management in system companies', *R&D Management*, **29** (1), 57–69.

Boutellier, R., O. Gassmann and M. von Zedtwitz (eds) (1999), *Managing Global Innovation*, Berlin: Springer.

Bresnahan, T. F. and M. Trajtenberg (1995), 'General purpose technologies: engines of growth?', *Journal of Econometrics*, **65** (1), 83–108.

Burgelman, R. A. (1994), 'Fading memories: a process theory of strategic exit in dynamic environments', *Administrative Science Quarterly*, **39** (1), 24–56.

Cheetham, P. S. J. (1999), 'The flavour and fragrance industry', in V. Moses, R. E. Cape and D. G. Springham (eds), *Biotechnology: The Science and the Business*, Amsterdam: Harwood Academic Publishers, pp. 533–62.

Chesbrough, H. (2001), '*Assembling the Elephant: A Review of Empirical Studies on the Impact of Technical Change upon Incumbents Firms*', Comparative Studies of Technological Evolution series, Amsterdam: JAI. Elsevier Science Ltd., pp. 1–36.

Cockburn, I., R. M. Henderson, L. Orsenigo and G. P. Pisano (1999), 'Pharmaceuticals and biotechnology', in D. C. Mowery (ed.), *US Industry in 2000: Studies in Competitive Performance*, Washington, DC: National Academy Press, pp. 363–98.

Coombs, R. and S. Metcalfe (2000), 'Organizing for innovation: co-ordinating distributed innovation capabilities', in N. J. Foss and V. Mahnke (eds), *Competence, Governance, and Entrepreneurship*, Oxford: Oxford University Press, pp. 209–31.

Dachs, B., T. Roediger-Schluga, C. Widhalm and A. Zartl (2001), *Mapping Evolutionary Economics – a Bibiometric Analysis*, paper prepared for the EMAEE 2001 Conference, Vienna, 13–15 September.

Daniell, E. (1999), 'Polymerase chain reaction: development of a novel technology in a corporate environment', in V. Moses, R. E. Cape and D.G. Springham (eds), *Biotechnology: The Science and the Business*, Amsterdam: Harwood Academic Publishers, pp. 147–54.

de Vos, W. M. (1999), 'Safe and sustainable systems for food-grade fermentations by genetically modified lactic acid bacteria', *International Journal of Economics and Business*, **9** (1), 3–10.

Delcour, J., T. Ferain and P. Hols (2000), 'Advances in the genetics of thermophilic lactic acid bacteria', *Current Opinion in Biotechnology*, **11** (5), 497–504.

Denrell, J., C. Fang and S. Winter (2003), 'The economics of strategic opportunity', *Strategic Management Journal*, **24** (10), 977–90.

Dugas, B., A. Mercenier, I. Lenoir-Wijnkoop, C. Arnaud, N. Dugas and E. Postaire (1999), 'Immunity and probiotics', *Trends in Immunology Today*, **20** (9), 387–90.

Ehrnberg, E., and N. Sjöberg (1995), 'Technological discontinuities, competition and firm performance', *Technology Analysis and Strategic Management*, **1** (7), 94–107.

Eliasson, G. (2000), 'Industrial policy, competence blocs and the role of science in economic development', *Journal of Evolutionary Economics*, **10** (1/2), 217–41.

Freeman, C. and L. Soete (1997), *The Economics of Industrial Innovation*, Cambridge, MA: MIT Press.

Gambardella, A. (1995), *Science and Innovation – The US Pharmaceutical Industry During the 1980s*, Cambridge: Cambridge University Press.

Granstrand, O., P. Patel and K. Pavitt (1997), 'Multi-technology corporations', *California Management Review*, **39** (4), 8–25.

Grupp, H. (1998), *Foundations of the Economics of Innovation*, Cheltenham, UK and Northampton, MA: Edward Elgar.

Henderson, R. M. (1993), 'Underinvestment and incompetence as responses to radical innovation: evidence from the photolithographic alignment equipment industry', *RAND Journal of Economics*, **24** (2), 248–71.

Henderson, R. M. and K. B. Clark (1990), 'Architectural innovation: the reconfiguration of existing product technologies and the failure of established firms', *Administrative Science Quarterly*, **35** (1), 9–30.

Investigational Drugs Database (IDdb) (2003), www.current-drugs.com/products/iddb/index.html

Jeffcoat, R. (1999), 'The impact of biotechnology on the food industry', in V. Moses, R. E. Cape and D. G. Springham (eds), *Biotechnology: the Science and the Business*, Amsterdam: Harwood Academic Publishers, pp. 515–32.

Judson, H. F. (1979), *The Eighth Day of Creation: Makers of the Revolution in Biology*, London: Penguin Books.

Konings, W. N., J. Kok, O. P. Kuipers and B. Poolman (2000), 'Lactic acid bacteria: the bug of the new millennium', *Current Opinion in Microbiology*, **3** (3), 276–82.

Kuipers, O. P. (2000), 'Genomics for food biotechnology: prospects of the use of high-throughput technologies for improvement of microorganisms', *Current Opinion in Biotechnology*, **10** (5), 511–16.

Kvistgaard, M. (1990), *Spredning af Bioteknologi til Dansk Erhvervsliv* [Diffusion of Biotechnology in Danish Industry], Copenhagen: TeknologiNævnet.

Leisner, J. (2002), 'Mælk og Bakterier' [Milk and bacteria], *Erhvervshistorisk Årbog*, 51.

Liebeskind, J. P., A. L. Oliver, L. G. Zucker and M. B. Brewer (1996), 'Social networks, learning and flexibility: sourcing scientific knowledge in new biotechnology firms', *Organization Science*, **7** (4), 428–43.

Lynskey, M. J. (2001), *Technological Distance, Spatial Distance and Sources of Knowledge: Japanese 'New Entrants' in 'New Biotechnology'*, Comparative Studies of Technological Evolution series, Amsterdam: JAI. Elsevier Science Ltd., pp. 127–205.

Margolis, J. and G. Duyk (1998), 'The emerging role of the genomics revolution in agricultural biotechnology', *Nature of Biotechnology*, **16** (4), 311.

Meyer-Krahmer, F. and U. Schmoch (1998), 'Science-based technologies: university–industry interactions in four fields', *Research Policy*, **27** (8), 835–51.

Miyazaki, K. (1994), 'Search, learning and accumulation of technological competencies: the case of optoelecronics', *Industrial and Corporate Change*, **3** (3), 631–54.

Morange, M. (1998), *A History of Molecular Biology*, Cambridge, MA: Harvard University Press.

Mowery, D. C., R. R. Nelson, B. N. Sampat and A. A. Ziedonis (2001), 'The growth of patenting and licensing by US universities: an assessment of the effects of the Bayh–Dole act of 1980', *Research Policy*, **30** (1), 99–119.

Noll, M., D. Fröhlich, A. Kopcsa and G. Seidler (2000), '*Knowledge in a picture*', Seibersdorf research report OEFZS-S-0101, November.

Orsenigo, L., F. Pammolli and M. Riccaboni (2001), 'Technological change and network dynamics. Lessons from the pharmaceutical industry', *Research Policy*, **30** (3), 485–508.

Pavitt, K. (1998), 'Technologies, products and organization in the innovating firm: what Adam Smith tells us and Joseph Schumpeter doesn't', *Industrial and Corporate Change*, **7** (3), 433–52.

Powell, W. W. (1998), 'Learning from collaboration: knowledge and networks in the biotechnology and pharmaceutical industries', *California Management Review*, **40** (3), 228–40.

Pridmore, R. D., D. Crouzzillat, C. Walker, S. Foley, R. Zink, M.-C. Zwahlen, H. Brüssow, V. Pétiard and B. Mollet (2000), 'Genomics, molecular genetics and the food industry', *Journal of Biotechnology*, **78** (3), 251–8.

Reid, G. (1999), 'The scientific basis for probiotic strains of *lactobacillus*', *Applied and Environmental Microbiology*, **65** (9), 3763–6.

Roberfroid, M. B. (2000), 'Prebiotics and probiotics: are they functional foods?', *American Journal of Clinical Nutrition*, **71** (suppl), 1682S–1687S.

Roseboom, J. and H. Rutten (1998), 'The transformation of the Dutch agricultural research system: an unfinished agenda', *World Development*, **26** (6), 1113–26.

Rosenberg, N. (1985), 'The commercial exploitation of science by American industry', in K. B. Clark, R. H. Hayes and C. Lorenz (eds), *The Uneasy Alliance: Managing the Productivity–Technology Dilemma*, Boston, MA: Harvard Business School Press, pp. 19–52.

Saarela, M., G. Mogensen, R. Fondén, J. Mättö and T. Mattila-Sandholm (2000), 'Probiotic bacteria: safety, functional and technological properties', *Journal of Biotechnology*, **84** (3), 197–215.

Salter, A., P. D'Este, K. Pavitt, P. Patel, A. Scott, B. Martin, A. Geuna and P. Nightingale (2000), *Talent, Not Technology: The Impact of Publicly Funded Research on Innovation in the UK*, SPRU, University of Sussex.

Senker, J. (1987), 'Food technology in retailing – the threat to manufacturers', *Chemistry and Industry*, **20** (July), 483–6.

Sharp, M. and J. Senker (1999), 'European biotechnology: learning and catching-up', in A. Gambardella and F. Malerba (eds), *The Organization of Economic Innovation in Europe*, Cambridge: Cambridge University Press.

Simon, H. A. (1996), *The Sciences of the Artificial*, Cambridge, MA: MIT Press.

Smith, K. (2001), 'What is the "knowledge economy"? Knowledge-intensive industries and distributed knowledge bases', paper presented to Danish Research Unit for Industrial Dynamics (DRUID) Summer Conference on The Learning Economy – Firms, Regions and Nation Specific Institutions Aalborg, Denmark, 15–17 June 2001.

Tannock, G. W. (1997), 'Probiotic properties of lactic-acid bacteria: plenty of scope for fundamental R&D', *Trends in Biotechnology*, **15** (7), 270–4.

Teece, D. J. (1986), 'Profiting from technological innovation: implications for integration, collaboration, licensing and public policy', *Research Policy*, **15** (6), 285–306.

Tushman, M. L. and P. Anderson (1986), 'Technological discontinuities and organizational environments', *Administrative Science Quarterly*, **31** (3), 439–65.

Tushman, M. L., P. Anderson and C. O'Reilly (1997), 'Technology cycles, innovation streams, and ambidextrous organizations: organization renewal through innovation streams and strategic change', in M.L. Tushman and P. Anderson (eds), *Managing Strategic Innovation and Change*, Oxford: Oxford University Press, pp. 3–23.

Unilever home page (2003), http://research.unilever.com/1_0/1_0_1-aboutus.html.

Utterback, J. M. (1994), *Mastering the Dynamics of Innovation*, Boston, MA: Harvard Business School Press.

Valentin, F. (2000), *Danske Virksomheders Brug af Offentlig Forskning en Casebaseret Undersøgelse* [The Use of Public Sector Research in Danish Firms], Copenhagen: Danmarks Forskningsråd.

Valentin, F. and R. L. Jensen (2002), 'Reaping the fruits of science', *Economic Systems Research*, **14** (4), 363–88.

Valentin, F. and R. L. Jensen (2004), 'Networks and technology systems in science-driven fields: the case of European biotechnology in food ingredients', in Jens Laage-Hellman, M. McKelvey and A. Rickne (eds), *The Economic Dynamics of Modern Biotechnologies: Europe in Global Trends*, Cheltenham, UK and Northampton, MA: Edward Elgar.

van der Meulen, B. and A. Rip (1998), 'Mediation in the Dutch science system', *Research Policy*, **27** (8), 757–69.

Wagner, D. R., C. Pierson, T. Warner, M. Dohnalek, J. Farmer, L. Roberts, M. Hilty and E. Balish (1997), 'Biotherapeutic effects of probiotic bacteria on candidiasis in immunodeficient mice', *Infection and Immunity*, **65** (10), 4165–72.

9. Commercialization of corporate science and the production of research articles

Robert J. W. Tijssen[1]

1. INTRODUCTION

This chapter is framed in the resource-based view of the firm. Among the many resource-related factors that influence a firm's organizational competitive advantages and business performance is its ability to innovate, to improve existing processes and products, and to produce new goods and services for the marketplace (Barney, 1991). The realization of the firm's primary role as a knowledge creator, as well as knowledge applicator, has led to knowledge-based theories of the firm (Grant, 1995), where R&D-intensive technology companies generate, accumulate and apply scientific and technical knowledge to produce incremental or breakthrough technological innovations. The intricate relationship between investments in scientific research, technological development, tacit knowledge resources, and technological innovations are generally recognized to be an important driver of competitive advantage of technology firms. There is empirical evidence that a firm's R&D efforts may directly improve its ability to innovate (Griliches, 1979), and indirectly help the firm to absorb outside knowledge (Cohen and Levinthal, 1990), both of which have a profound impact on the firm's productivity (Hall, 1996). The research base is acknowledged to be a critical element of a firm's innovation capability through the (semi-)open and continuous interaction that takes place with external information sources, such as universities and other public research institutions. Empirical studies have shown that many corporate technical inventions and related innovations depend upon scientific progress (Beise and Stahl, 1999; Mansfield, 1991; Tijssen, 2002).

Traditionally, the creation of scientific knowledge and associated technical knowhow was viewed as a linear process in which firms endogenously seek out and apply these knowledge inputs, in the form of R&D efforts, to generate commercially valuable innovative output. Recent developments in evolutionary economics view this process as the outcome of context-specific and firm-specific learning processes that bring in their preexisting competencies, experience and knowledge (David and Foray, 1995). The linkages and interactions among the economic agents who produce, diffuse and adopt this knowledge are seen as crucial for the commercialization of knowledge inputs into innovative outputs. Owing to this interactive and semi-open knowledge creation system, the same firms that produce the knowledge do not always appropriate the expected returns of their R&D efforts. Voluntary knowledge transfers or involuntary spillovers of scientific and technical knowledge may be absorbed and utilized by other firms, especially in the case of exploratory scientific and technical research. R&D-intensive firms therefore face the problem of effectively reconciling the production, protection and dissemination of their research-based knowledge.

This chapter presents the results of an empirical study of worldwide quantitative data on corporate research outputs. The results suggest that the balance is shifting in favour of knowledge protection and appropriation, rather than production and dissemination. Section 2 elaborates on the relevant concepts and economic issues associated with this 'knowledge flow balance' in the corporate research sector. Section 3 describes the main features of the information sources and analysis methods. The main findings are presented in Section 4, followed by tentative conclusions and some cautionary remarks in Section 5.

2.   KNOWLEDGE APPROPRIATION AND KNOWLEDGE SPILLOVERS

2.1   Corporate Basic Research: Life Blood or Bleeder?

After the golden age in the 1960s and 1970s, and following the cutbacks and the short lived upswing in corporate science spending in the late 1980s and the early 1990s (e.g. Rosenberg, 1990), a gradual reorientation of business strategies and IPR policies took off in the mid 1990s when industrial research labs became 'leaner and meaner'. Labs became smaller, more decentralized, and their scientific and business performance more closely linked to corporate strategic planning and investor confidence. Many of those structural changes started either in Japan or the USA. Researchers and engineers in the labs are now made accountable for their actions and outputs, including material for research papers in the journal literature (Buderi, 2000a; Varma, 2000). Further empirical evidence indicates that this evolution in the industrial research landscape is still ongoing: companies have prioritized R&D to stay competitive in the long run, but at the same time many large firms have downsized their central research labs (Coombs and Georghiou, 2002). As a result, corporate research now seems to be more than ever driven by new business creation and the commercialization of research outcomes into marketable products, processes and services. As the introduction of new products has vastly increased and lead times have diminished, corporate research endeavours are now also continuously assessed and evaluated in terms of quality, productivity and (potential for) value creation.

Faced with increasing competition and shorter development cycles, companies' innovation strategies include collaboration with major sources of new knowledge creation around the world. This is most marked in the new research-intensive sectors where the underlying science is extremely dynamic, the technologies are strongly science-related and the development sometimes takes place at the interface between different disciplines and fields. Knowledge derived from scientific research or engineering research provides an invaluable understanding and theoretical base for innovation-oriented corporate R&D activities. The benefits of research are derived not only from in-house applied research, but also from basic research both internal and external (Griliches, 1986; Mansfield, 1981). Obviously, the business sector has always been engaged in scientific research primarily out of self interest. Basic research is usually a costly activity with uncertain strategic benefits or monetary gains. Longer term research has therefore always been a small part of corporate R&D – on average some 10 per cent of the business R&D expenditures are devoted to research with a long term orientation. This kind of research is traditionally confined to the large R&D-intensive companies and their central laboratories. An increasingly large share of the funding for these laboratories now comes from business groups and product divisions through contractual agreements about programs and costs, a fraction ranging up to an estimated 75 per cent in the case of the large US companies (Larson, 2001). Since many of these science-intensive companies are now operating in rapidly changing technology areas and markets, the term 'basic research' is often no longer appropriate.[2] Long term research projects are often reduced to schedules of no more than two or three years. Corporate research portfolios now balance short term deliverables with long term objectives, where each of the largest R&D-intensive firms has developed various organizational and management models to balance the needs of business groups and the R&D agenda of the central research laboratories (Buderi, 2000b).

This process of the 'marketization' of research places a stronger emphasis on the protection of research findings and on the exploitation of intellectual property rights, especially in the case of findings of (potential) commercial or strategic value. Once scientific or technical information is disseminated intentionally (or spilt over unintentionally) into the public domain competitors are free to benefit. One may argue that firms ought to perform only the most essential basic research, closely guard valuable results, and commercialize it rather than making it public through publications or otherwise. Hence, why do firms still bother to invest in longer term research when those large investments and commercial risks may outweigh business advantages? More importantly, why should corporate researchers still want to publish their findings in the open literature? Providing answers to these questions requires further examination of the economic relevance of corporate knowledge bases, the role of scientific and technical research in absorbing relevant knowledge, and intricate links between knowledge appropriation strategies and dissemination practices.

2.2   Knowledge Bases and Absorptive Capacity

Mainstream economic theory considers scientific knowledge as a uniformly available public good that can be transferred and learnt at little cost. Nonappropriability and indivisibility of knowledge, coupled with inherent uncertainty as to the results of basic research, were supposed to lead to underinvestment by firms in basic science (Arrow, 1962; Nelson, 1959). True enough, cutting-edge basic research tends to be expensive and risky, and once results are written down ('codified') and made publicly available, every firm can enjoy the knowledge freely. Firms may therefore be reluctant to invest because knowledge spills over easily from first movers and innovating firms to other firms that can free ride on the efforts of the innovators. Hence, firms have relatively little incentive to do basic research, or so it is argued. Several researchers have questioned whether spillovers of valuable knowledge occur as easily as portrayed by Nelson and Arrow. Economic researchers in the 1980s and 1990s have challenged these assumptions and concluded that firms require an appropriate knowledge base, and need to perform their own basic research in order to absorb and appropriate 'free' scientific information and technical knowhow (Mowery, 1983).[3]

Knowledge creation based on scientific and technical information is a complex and cumulative process. To absorb and assimilate codified scientific and technical knowledge requires a certain measure of learning by doing at appropriate levels of tacit knowledge (Nelson, 1989; Pavitt, 1991). Firms need a research base – either in-house or externally – which covers all resources from which new scientific and engineering knowledge can be drawn. The intensity and effectiveness of these research-related interactions are determined to a large extent by the firms' own commitment to learning activities and the ability of firms to recognize and appreciate the value of new, external information (ranging from generic science to new production equipment), to assimilate it, and exploit its economic potential through commercialization. Cohen and Levinthal (1989, 1990) have labelled this ability as the 'absorptive capacity' of the firm, which largely depends on the level of prior related knowledge owned by the firm. Cohen and Levinthal argue that when learning is especially difficult, a firm's ability to apply external basic research for its own commercial gain is a function of its R&D investment. Given that learning is a highly localized and history-dependent process, the current set of skills and expertise owned by a firm is critical for the nature and direction of learning processes that aim to enhance the knowledge base of the firm in the future. Following this argument, the ability of a firm to use the results of research efforts made by other firms or other public research organizations depends on its ability to understand them and to assess their economic potential; an ability affected by the size of a company and its access to complementary assets (Levin et al., 1987; Teece, 1987). Thus, lack of tangible or intangible investments (in the form of human capital) in an area of expertise early on may inhibit the development of technological knowledge and innovations by the firm in that area at a later stage.

2.3 Cooperation, Networking and Knowledge Flows

The most effective way for companies to evaluate and monitor key outcomes of their in-house research, and assess the potential of external research, is to participate in research communities – either local, domestic, or on the global level. However, the private sector engages in scientific research only when expected private returns rise above a minimum level. Moreover, in-house R&D efforts alone might no longer be able to create enough economic value to warrant large expenditures on basic research given the uncertain outcomes of exploratory scientific and engineering research. In order to reap economies of scale and scope beyond the reach of a single company, or to offset their own slimmed down corporate research programs, many large R&D-intensive technology firms have increasingly turned to outsourcing and subcontracting long term research to universities, or are now engaged in joint research ventures, thus creating a new 'industrial ecology' of corporate R&D (Coombs and Georghiou, 2002), where public sector research has become a major source of new scientific knowledge and advanced technical skills through various transfer channels (Salter and Martin, 2001). This pervasive development can be seen in the rise of institutionalized cooperative structures such as interfirm joint research ventures and university–industry strategic research partnerships (Hagedoorn et al., 2000). Analysts claim that many large R&D-based companies have forged stronger informal relationships and formal (contract-based) linkages with public sector research organizations, and industry now relies largely on universities and research institutes to explore new avenues of research and generate new knowledge (Meyer-Krahmer and Schmoch, 1998). OECD data suggests that a large share of corporate funding for basic research is being spent on joint ventures with external research partners, in particular within the local or domestic university sector.[4] Firms that are in the business of developing new products are more likely to find public research to be an important source of information than firms that innovate in order to improve their existing products (Arundel and Garrelfs, 1997). Indeed, several large pharmaceuticals companies, like Pharmacia, Syngenta and Amgen, have closed multimillion dollar deals in the late 1990s with US universities in order to obtain (exclusive) access to results of frontier basic research.

The nature of the knowledge generation process itself seems to be evolving towards a more network-embedded process with a stronger emphasis on the interplay between knowledge demand and knowledge supply, as well as increased levels of transdisciplinarity of the research projects, and a larger degree of heterogeneity of the actors involved (Gibbons et al., 1994). By transforming industrial R&D problems into research topics for basic research, many public–private research partnerships and joint research ventures benefit the research agendas of academic researchers and engineers. The connection to academic science and scientific networks shapes both scientific advances and technological progress.

The emphasis on access to external sources, and the view of knowledge as the outcome of interactive learning processes, implies the existence of knowledge flows that link different sources of new scientific and technological information and its potential users. These flows of knowhow and information include voluntary dissemination, intentional transfers, as well as accidental or unintended spillovers. There are many communication channels and routes for these knowledge flows to materialize. For example, Cockburn and Henderson (1998), building on Cohen and Levinthal's notion of absorptive capacity, suggest that the degree to which firms are connected to universities is an important factor for utilizing those external knowledge flows. The general findings from their research suggest that three major types of connections and modes for knowledge flows exist: (1) research publications and co-authorships; (2) proximity to star scientists; and (3) human resource movements. In the following section we will focus on the linkages and flows embodied in research publications to describe recent worldwide output trends within an increasingly 'networked' system of corporate basic research.

2.4 Corporate Research Papers in the Open Literature

With regards to codified knowledge resulting from corporate basic research, there exists a natural tension between sharing and protecting that information. Publishing research findings in scientific literature is at odds with the industry's tendency to privatize scientific and technological knowledge. In many instances it is in the self interest of industrial actors not to publish since that could be beneficial for their competitors. Industrial researchers and engineers would seem to have little incentive to disseminate the results of proprietary research that might be interesting and helpful to outsiders. However, each year many thousands of research documents related to, or directly originating from corporate research, become public knowledge. More than 10 000 research articles are published annually in international scientific and technical journals (see Section 4). Clearly, writing publications for these peer-reviewed journals costs valuable time and corporate money whereas the commercial benefits would seem uncertain or are sometimes marginal at best. So why do corporate research labs still act as quasi-academic research labs and publish these large quantities of papers?

Nelson (1990) argues that firms have many good reasons to publish (selected) results of their research endeavours of low competitive value, in order to not only maximize visibility and link up to the scientific community, but also to establish intellectual claims and legal rights, and to signal R&D capabilities to (potential) partners and suppliers.[5] Additional incentives of more recent data include R&D management objectives such as: attracting private capital and public research funding, and gaining a reputation and enhanced credibility for doing high quality (basic) research which attracts first rate researchers and technicians. As such, publications not only represent the firm's production of scientific and technical knowledge as a public good, but also acts as a gateway in a two directional knowledge transfer pathway between the firm and scientific communities in the outside world: in the case of low rivalry conditions, firms may expect reciprocity exchange effects where publications may induce further research by others.[6] This 'open science' mechanism produces a pool of knowledge that can be used freely by the international scientific community from which corporate researchers draw very heavily (Jaffe, 1989).

Obviously, companies will publish only a fraction of their research findings that are of interest to the relevant scientific and engineering research communities. Firms will carefully balance their desire for secrecy and their willingness to share and disseminate information. It will usually be decided to keep (potentially) valuable information in-house, and only in exceptional cases will a more open publication policy be enforced in which some key information is shared or exchanged. The latter will involve the screening of manuscripts and partial dissemination of research results, imposing strict conditions on accessing research material and outcomes, and enforcing delays in publication. Given the strategic nature of corporate research and the importance of intellectual property rights, these publications should mainly be seen in the light of corporate business strategies. Zucker et al. (1998) asked how a firm's linkages to scientific networks affects its overall economic performance and more specifically its technological progress, particularly in instances when novel technologies are science-based. With respect to the role of research publications in these linkages, it has been argued that particularly in periods when there is a shift in the technological paradigm to one closely linked to science, publications by the leading firms are crucial for mobilizing relevant in-house research and external research to make a successful transition.

Whatever explains corporate scientific publishing, it is obvious that publishing is not the main purpose of corporate researchers and engineers, and firms publish many fewer research articles than comparable public sector institutions (universities, research institutes and government laboratories) with the same research resources and working in the same fields of science. Moreover, if firms do decide to publish, many of these papers are likely to be co-authored with researchers in the public sector. In the case of joint research partnerships with public sector organizations, the corporate sector is bound to apply slightly different knowledge management considerations and strategies in view of dissemination-oriented research missions, their incentive structures, and intellectual property rights (IPR) policies of their partners in the public sector. Corporate sponsors of public research engaged in contract-based ('formal') cooperation will often negotiate the first rights (of refusal) to the fruits of research and the scientists must delay publishing to allow companies a head start for commercializing through filing for patents by other means. Scientific cooperation with public research organizations on a more 'informal' personal basis is more likely to generate jointly authored research papers, especially in the case of academic partners who have strong incentives to publish results related to research sponsored by industry, or conducted in cooperation with the corporate sector. Irrespective of the nature of contractual agreements, in the process of producing these co-authored scientific papers, researchers are likely to exchange tacit and embodied elements of knowledge and skills. These co-authored research papers therefore not only gauge the production of new collective knowledge, but also the absorption of external knowledge by the firm during knowledge creation and codification.

In most areas of international open science, the main channel of disclosure of codified knowledge is that of conference proceedings, or research articles published in the quality-controlled peer-reviewed international

scientific and technical journals. The next section turns to the further introduction of the latter type of research publication and related measurement
issues.

3.   INFORMATION SOURCES AND METHODOLOGY

3.1   Bibliometric Analysis of Corporate Basic Research

General trends in the output of basic research efforts within large science-
intensive technology firms, or for that matter entire science-based industries,
can be derived from statistical analyses of the quantity of papers published in international peer-reviewed scientific and technical journals. This
literature-based ('bibliometric') approach produces a large body of quantitative data that provides a statistically robust frame of reference for
analysing the changing contribution of corporate research in research communities. Sets of research papers originating from the same (parent) company
enable comparisons between firms, and the aggregation of those firm-level
data allows for comparisons between associated science areas and industrial
sectors. The number of co-authored papers originating from (informal) joint
research ventures – intrafirm, interfirm and public–private – enable a range of
statistical analyses on the volume and composition of cooperative corporate
basic research. Although these joint papers are considered useful proxies of
these cooperation-based knowledge flows and exchange, they should be
handled with due care as a reliable source of conclusive empirical evidence
on actual scientific cooperation (Katz and Martin, 1997).

Bibliometric studies of corporate publication output in international
journals conducted as early as the 1970s have provided empirical data
on trends in the 1980s up until the mid 1990s, especially for US industry
(e.g. Halperin and Chakrabarti, 1987; Small and Greenlee, 1977). The findings revealed significant increases in the 1980s and early 1990s, resulting in
a 5–10 per cent share of the corporate sector in the global scientific output.
Several studies have focused on large firms, a single industry, or the distribution over papers across industrial sectors within one country (e.g. Godin,
1996; Hicks et al., 1994; Hicks and Katz, 1997; Tijssen et al., 1996). However,
to our knowledge no systemic study has been made of worldwide output
levels and trends across all sectors and countries.

Returning to the major socioeconomic forces impacting upon basic
research and publication strategies of modern day corporate researchers
(i.e. the three 'C's: Competitiveness, Cooperation, and Commercialization),
the aggregate-level bibliometric data on their research papers in international journals allow us to address the following key questions:

1. To what extent have the competitive pressures in the 1990s forced science-based industries to commercialize their research efforts and to shift their focus from being a 'science performing industry' towards operating as a 'science using industry'? More specifically, has the published research output of the corporate sector dropped, and has the number of co-authored research papers increased at the same time?
2. How has this reorientation impacted on cooperative research ventures of firms, especially those with other firms – as opposed to partnerships with public sector research institutes and universities? In other words, have the share and composition of jointly authored corporate research papers changed?
3. And to what extent are the observed trends universal or sector-specific? Do we find different trends in the major science-based industries?

3.2 Databases and Definitions

Providing answers to the above questions requires a comprehensive database of corporate research papers covering all relevant industrial fields of science, and the major research-based firms and industrial sectors. The bibliometric study is restricted to the internationally visible production of corporate research papers covered by the large multidisciplinary bibliographic databases compiled by Thomson–ISI. These ISI databases, especially the *Science Citation Index®*, provides the best source of information to identify basic research activity across all countries and all fields of science. The statistical analyses were done with CWTS's tailored version of the ISI databases. The research papers include all document types that, in varying degrees, originate from original basic research: research articles, review articles, research notes, and letters (editorials, book reviews, etc. are omitted). The vast majority of those papers are research articles. CWTS assigns each paper only to those (main) institutions where the address information refers unmistakably to the respective (main) organization(s).

The analysis covers all research papers listing at least one author affiliate address referring to an organization that CWTS classified as being part of the 'corporate sector'. The demarcation of this sector is based on the following general definition: all business enterprises, organizations and institutions whose primary activity is the commercial production of goods and services (other than higher education and medical care) for sale to the general public at an economically significant price. This institutional delineation includes public–private consortia, private nonprofit and not-for-profit institutions and government-owned nonprofit companies. Also included are private nonprofit R&D organizations mainly serving the business enterprise sector, or privately funded research institutes and other

R&D performing institutions (other than the higher education sector or the medical care sector, and/or mainly controlled by and funded by government).

Data cleaning, unification, and consolidation of those papers to parent companies was done using information on websites and occasionally Dunn & Bradstreet's *Linkages* database (formerly the *Who Owns Whom* Directory of Corporate Affiliations database). The data were consolidated in mid 2002, in most cases at the 'main organizational' level of the legal entity (i.e. parent companies, R&D labs, universities, research institutes, etc.). Corporate research laboratories, majority-owned subsidiaries and other corporate affiliations are included as far as possible in the current parent company. Companies added to the parent through mergers and acquisitions in the years 1996–2002 were renamed to the current parent company to ensure compatibility over time. In the case of multinational companies, a consolidated group name is defined which refers to the ultimate parent (or holding). Foreign branches and foreign subsidiaries of a company are labelled with the same consolidated name, or are listed by their country-specific consolidated name. Each (parent) company is linked to the country of location mentioned in the author address.

Counts of co-authored papers are defined at the level of these main organizations.[7] Each organization is defined at the highest aggregate level – the main organizational level. They are assigned to the country of location as listed in the affiliate address on the research publications. Dividing up a paper between the participating units (researchers, organizations, countries) is to some extent arbitrary – there is no fair method to determine how much money, effort, equipment and expertise each entity contributes to the underlying research effort and writing the paper. Our basic assumption therefore is that each author, and associated organization, made a non-negligible contribution. Consequently, we adopt a counting scheme in which each paper is fully allocated to each of the main organizations listed in the author address heading.

It is important to stress that an unknown fraction of the corporate research papers are probably not exclusively basic research-oriented; they will also relate to application-oriented ('strategic') research as well, and perhaps to a certain degree also 'applied' research that is directly related to technological development. Since universities are generally accepted to be the major locus of curiosity-driven 'blue sky' basic scientific research, corporate papers listing at least one university are assumed to be more basic research-oriented as compared to co-publications listing nonuniversity public sector research organizations. The papers jointly authored with nonuniversity research organizations, and especially those with other firms, are assumed to represent strategic research rather than basic research.

The publication output analyses distinguishes six types of corporate research papers; in addition to papers that were authored solely by one private sector organization, we define the following mutually exhaustive set of categories of jointly authored research papers:

1. Two firms exclusively;
2. Three or more firms exclusively;
3. One firm with public sector organizations – including one or more universities;
4. One firm with public sector organizations – excluding universities;
5. Two or more firms with public sector organizations – including one or more universities;
6. Two or more firms with public sector organizations – excluding universities.

3.3 Industrial Sectors

Sector-level analyses deal with two R&D-intensive high technology industrial sectors: (1) pharmaceuticals, and (2) the semiconductors industry. Both are characterized by strong relationships between research, technological development, and innovation. Both involve difficult learning environments where research-based scientific and technical knowledge play an important role in knowledge creation and exploitation. Basic research in the pharmaceuticals industry explores the genetic and biomolecular mechanisms of diseases in relation to designs of drugs. For semiconductors, basic research includes the physics of solid state devices and the chemistry involved in manufacturing integrated circuits. Corporate in-house longer term research plays a stronger role in the pharmaceuticals industry, where a firm's progress and competitive position are closely tied to advances in basic research and knowledge appropriation through patenting. The expected benefits of basic research for design of drugs are therefore much higher than, for example, the design of new materials for semiconductors.

These sectors are defined in terms of a representative set of firms that were selected from two public databases previously or currently available on the Internet:

1. 'R&D Scoreboard 2001' compiled by the UK Department of Trade which covers the annual accounts of the 500 largest R&D spenders worldwide in the period 1996/97–2000/2001 (www.innovation.gov.uk/projects/rd_scoreboard/database/);
2. 'TR Patent Scorecard 2002', a joint effort of Technology Review and CHI Research, Inc., covering firm-level R&D performance data based

on CHI's analyses of their USPTO patents granted in 1996–2001 (www.technologyreview.com/scorecards/patent_2002.asp).

A joining of both databases for the two industrial sectors resulted in the following sets of companies, which include most of the large and scientifically leading firms across the globe:

1. Pharmaceuticals: 87 firms (55 North America, 16 Asia, 16 Europe);
2. Semiconductors: 75 firms (51 North America, 21 Asia, 3 Europe).

The two lists of companies, and their countries of headquarters, are presented in Tables 9.1 and 9.2. Each set includes those firms that published at least one research paper during the period 1996–2001 indexed within the ISI/CWTS database. Note that these sets are assumed to be representative only for the large R&D-intensive companies in these sectors, and not necessarily so for the sector as a whole, which includes many high tech start-ups, SMEs, and diversified companies classified in different primary business sectors. Nonetheless, given the number of selected companies and the fact that the selection includes the main R&D actors in these industries, as well as being the main contributors of research papers in international journals, we expect a reasonable coverage of published basic research outputs of the entire industrial sector.

4.   RESULTS OF THE ANALYSES

4.1   Diverging R&D Output Trends

To what extent are recent shifts in the marketization of industrial R&D, as described in Section 2, visible in the corporate R&D literature? If the major business enterprises in the advanced industrialized countries indeed spent the same amounts on basic research in the 1990s, but have become more focused on strategic/applied research rather than basic research, and now promote the protection and exploitation of science-based knowledge rather than dissemination in the open literature, we should expect to find at least some of the following trends in the available empirical data: (1) more corporate researchers; (2) declining budgets for basic research; (3) more patents – especially science-based patents; (4) fewer research papers in the international scientific literature; (5) less research cooperation with other companies, and (6) more cooperative linkages with universities and other public sector research organizations.

Worldwide some 290 000 articles were published in the period 1996–2001. The total publication output by the corporate sector shows a 12 per cent decrease during the interval 1996–2001 and this annual decline has accelerated in recent years (4 per cent in 2000; 10 per cent in 2001). As for research inputs, according to OECD figures its member states spent on average about 0.4 per cent of their GDP on basic research in the mid to late 1990s (OECD, 2001). However, country-level data on the share of the business sector are lacking, or difficult to compare, often due to shortcomings in the somewhat ambiguous concept 'basic research' as defined by OECD's *Frascati Manual* (Geullec, 2001).[8] More detailed information exists for only a few countries, including the USA where the business sector itself collects the data (Larson, 2001).[9] Fortunately, the Organisation for Economic Co-operation and Development (OECD) provides more comprehensive data on the quantity of researchers in the business sector. Using the trends on the total number of researchers in the OECD member states in the years 1994–99 as a baseline, we can examine various R&D output trends over the period 1996–2000/2001. The results are presented in Figure 9.1.

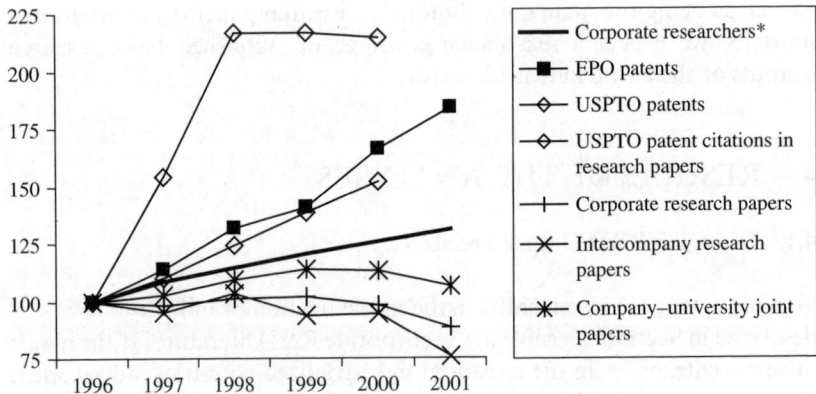

Notes:
*We assume a two year time lag between trends in volume of researchers and R&D outputs published in the open literature. The numbers of researchers in the business sector within the OECD refer to the period 1994–99 but are superimposed on the 1996–2001 axis for ease of comparison.

Data sources: USPTO *US Patent Statistics Report* – Summary table; *US Science and Engineering Indicators 2002*; EPO *Annual Reports* 2000 and 2001; ISI/CWTS database; OECD, MSTI database November 2001.

Figure 9.1 Diverging R&D output trends worldwide (1996 = 100)

The longitudinal analysis shows steadily increasing numbers of corporate researchers (an unknown fraction of which are involved in basic scientific and engineering research) in conjunction with a divergence in the output trends between the two major classes of codified R&D information: large growth rates of IPR-protected patents versus a gradual decline of the freely disseminated research papers in the journal literature. Moreover, we observe a significant growth rate in patent citations of the scientific literature, which corroborates the observed emphasis on the commercialization of science-based industrial R&D.[10] The divergence between both types of R&D output is fairly recent: the decline of corporate publishing has been a very gradual process up until 2000. Given the average time lag between research inputs and published outputs, this bifurcation process must have started in the mid 1990s, which seems to coincide with anecdotal evidence from other sources (see Section 2.1).

The volume of interfirm co-publications has deteriorated by 25 per cent since 1996, while the numbers of industry/university co-authored articles has gradually fallen back to the 1996 level. So, it would appear that one of the main factors driving the declining publication output relates to whether or not research partners are involved in corporate basic research, and the type of partners involved. Figure 9.2 exhibits a further breakdown of the trends in the various categories of co-authored research papers, as well as temporal changes in the numbers of single company-authored articles. The largest decline occurs for research papers listing only one company. We find an accelerating rate of decline in which the share of these papers has dropped significantly from 36 per cent to 26 per cent between 1996 and 2001. Co-publications involving pairs of companies are also in rapid decline. However, the drop in papers originating from research partnerships involving three or more firms is smaller than for pairs, suggesting different knowledge creation processes and appropriation regimes in corporate research partnering depending on the number of firms involved. It would seem that the larger the number of partners involved, the more the research will be of a generic 'pre-competitive' nature and the results are likely to be (partially) transferred to the open literature for strategic reasons.[11]

When universities are engaged in research partnerships with industry they act mainly as the producers of basic knowledge and advanced technical skills (and associated human capital in the form of PhD students and researchers), while corporate research partners focus on the transfer, absorption and assimilation of that knowledge and knowhow. This relatively clear cut division of labour and responsibilities, in conjunction with the industry's never ending need for new inputs of leading edge scientific knowledge, instruments and skills, ensures a fairly stable quantity of joint research papers with academics. The quantity of industry–university co-authored publications

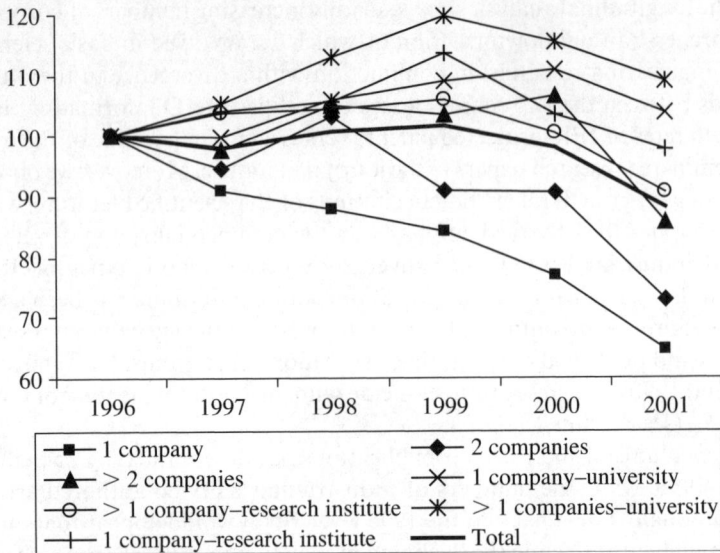

Sources for company selection: 'TR Patent Scorecard 2002' (Technology Review and CHI Research, Inc.); 'R&D Scoreboard 2001' (UK Department of Trade).

Data source: ISI/CWTS database.

Figure 9.2   Trends in corporate research articles worldwide, all industrial sectors (1996 = 100)

showed a 13 per cent gain in 2000, which slipped back to 6 per cent in 2001. Due to the larger rates of decline of the other categories of corporate papers, the fraction of these articles in the corporate output has increased steadily from 48 per cent in 1996 to 58 per cent in 2001. Coupling the industry's increased need for research cooperation with universities, and the output rewarding incentive systems in the academic community, would seem to ensure a sustained flow of joint research papers reflecting knowledge flows to firms for their in-house research, technological development and further commercial use.

Interestingly, industry–university co-publications involving multiple firms are less affected by the general downturn compared to single company co-publications with universities. As the size and heterogeneity of public–private research alliances and networks grows, especially those aimed at producing scientific or technical knowledge to be shared amongst all (major) partners, the more prone these partnerships seem to be to disseminate this research information into the public domain – not in the least to satisfy the researchers in the public sector who need to comply to publication output-driven rewards systems (e.g. Tijssen, 1998). In contrast, the volume of joint papers involving research institutes, or other nonacademic partners in the public sector, does show a noticeable decline from 1996. Since nonuniversity public sector researchers are less active in basic research and are less driven by publishing papers in international journals, the number of joint papers co-authored with corporate researchers are now also decreasing significantly. Overall, we see a pattern, similar to the trend found in the interfirm partnerships, where the number of partners involved in public–private co-publications is inversely correlated with the rate of decline.

4.2 Research Output Trends by Industrial Sector

Obviously, the overall trends depicted in Figures 9.1 and 9.2 hide a high degree of variation, both at the firm level and across different industrial sectors. It stands to reason that the underlying (changes in) volume of basic research and/or decreasing publication activity will vary by industry. A recent bibliometric study by Lim (2001), using research articles in international journals and USPTO patents, indicates a strong link between both outputs in the pharmaceuticals sector but a weaker relationship in the semiconductors industry. Lim argues that these differences are due to sector-level differences in the relevance of basic research for innovations in conjunction with firm-level differences in absorptive capacity of knowledge spillovers.

Figure 9.3 exhibits the breakdown of the various types of corporate research papers for both sectors, disclosing some sector-specific characteristics and developments. The large pharmaceuticals companies produced a staggering 55 962 papers in the period 1996–2001, displaying a remarkably high propensity to produce multiple-company papers, the number of which remained fairly stable in the years 1996–2001. This would seem to indicate a sustained tendency on the part of these (large) firms to take part in inter-firm or intrafirm research alliances.[12]

The large semiconductors companies, producing a total of 15 641 papers, exhibit a very large growth in the number of papers from partnerships involving one company and several nonuniversity public research organizations. The number of these papers rose by some 30 per cent since 1996. Furthermore, the quantity of papers listing several companies and one or more research institutes remains stable. This sector-specific finding ties in with the results of Lim (2001) suggesting that semiconductor firms depend primarily on applied knowledge rather than basic knowledge. In the semiconductors industry many intermediate steps are required to transform basic scientific breakthroughs into useful innovations, which reduces their

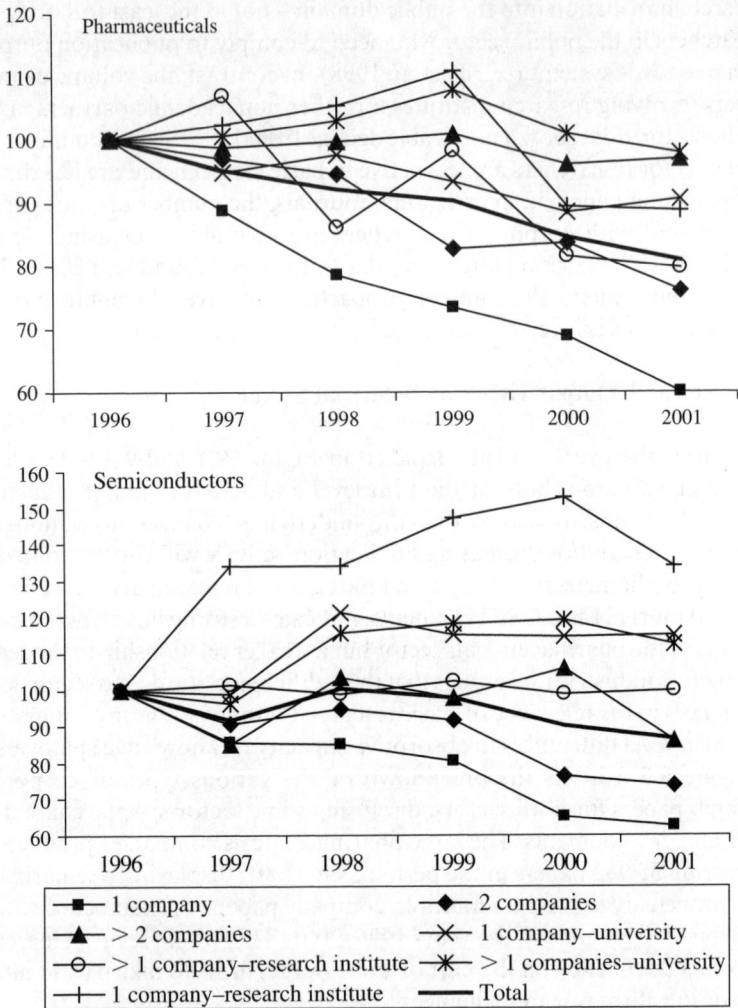

Sources for company selection: 'TR Patent Scorecard 2002' (Technology Review and CHI Research, Inc.); 'R&D Scoreboard 2001' (UK Department of Trade).

Data source: ISI/CWTS database.

Figure 9.3   *Trends in corporate research articles worldwide, selected industrial sectors (1996 = 100)*

need to invest heavily in their own longer term research. R&D-based semi-conductor firms seem to have increased their investments in precompetitive research at research institutes in the public sector, rather than boosting in-house basic or strategic research or engaging in strategic research ventures with other firms. The technology-oriented research institutes in the public sector, rather than general universities or technical universities, are sought out by the semiconductors industry as the main sources of applied scientific knowledge.

Similarly to the overall picture in Figure 9.2, Figure 9.3 may hide marked differences between the large firms, especially between science-based first moving 'innovators' and the 'followers' that spend less of their resources on R&D and in particular on basic research. However, given the lack of firm-level data, we can only assume that in view of the increasing international competitive pressures in both sectors, and the business practices shared by the major firms, there is no compelling reason to believe that large companies, active in the same fields and competing in the same local or global markets, are adopting fundamentally different strategies for enhancing the commercial pay-offs of their research efforts.

5. CONCLUDING REMARKS: IS CORPORATE RESEARCH IN DECLINE? OR ARE FIRMS JUST PUBLISHING LESS?

The erosion of the industry's contribution to the open scientific and technical literature gained momentum toward the turn of the millennium. We might be tempted to conclude that this trend follows entirely from firms switching their priorities to short term research focused on areas close to the market where they can make money more quickly. These competitive pressures to increase private rates of return, and to boost commercialization of research findings, may have redirected the goals of basic research and narrowed the focus towards strategic and applied research with shorter time horizons. Most likely, companies were also trying to minimize research costs by contracting out for work rather than conducting in-house research. Less funding for in-house exploratory research, and the downsizing of corporate research labs, would indeed account for the significant decrease of corporate research articles in the open literature, especially the dramatic decline in the publication rates of papers where companies are the sole creator of new scientific knowledge, as well as the significant drops in interfirm co-publications. Moreover, the relatively minor effect on industry–university papers can be explained by closer links with the university sector.

The downturn in corporate spending on basic research would also account for the significant differences we observe between output trends in joint papers with two partners and those listing three or more. Assuming that the papers listing many research partners arise from joint ventures and consortia that are primarily engaged with precompetitive research of a more generic nature, these partners have less reason to appropriate collective knowledge and impose more restrictive publishing strategies. In other words, in the case of basic research involving many partners, either in the corporate sector or public sector, knowledge dissemination practices tend to be less vulnerable to changes in corporate research culture. As for the industry's links with nonacademic research organizations in the public sector, it is safe to assume that these partnerships are more focused on strategic or applied research and therefore less affected by withdrawals from basic science. Moreover, publishing findings from this kind of joint research is likely to be more severely constrained by IPR arrangements and publication strategies compared to industry–university co-authored papers in view of the perceived commercial value of such research findings and the greater risk of unintended spillovers.

However, other organizational and socioeconomic factors might also (partially) explain these changes within corporate research culture and its effect on the propensity to publish in peer-reviewed journals. As state of art research has become more complex and expensive, and driven by tighter time schedules, research projects have become subject to stricter 'costs and returns' accounting rules that focus on milestones, tangible deliverables and value creation. As a result, the production of research articles has decreased because investing time and effort in writing these papers has become increasingly prohibitive, while other performance targets and R&D results such as patents and patent-based licenses are more highly rewarded and generate greater in-house recognition and reputation. Researchers may have gradually opted for other 'easier' publication outlets with less severe refereeing, such as internal report series, contributions in conference proceedings, or papers in professional journals.[13] Only the truly high quality papers are still submitted to peer-reviewed international journals (where related research papers, or earlier abridged versions of the same paper are made public through other outlets).

Concluding, the whole pattern of observations point in the direction of structural changes in corporate priorities and strategies concerning basic research leading to codified knowledge that is – in principle – publishable in the open scientific and technical literature. The observed trends in output of corporate research articles in recent years suggest, on the one hand, that corporate basic research is being downsized, but on the other hand that corporate priorities concerning access to their research-based knowledge, and securing related intellectual property rights, are probably also getting the better of sharing and exchanging information with the worldwide scientific community. The correlation and causality between diminishing resources for corporate basic research and the declining output levels require further investigation.

Nonetheless, based on the findings presented in this chapter we cannot rule out the possibility that science-based companies might still be doing the same magnitude of long term research, but that their R&D labs and research managers now operate in different organizational and managerial structures that are governed by rules and regulations aimed at maximizing the efficiency of knowledge creation processes and broadening the opportunities for commercial gains of research activities. Coupled with IPR-driven knowledge appropriation regimes and more restrictive policies for the dissemination of findings of general scientific importance, makes researchers shy away from publishing in peer-reviewed scientific and technical journals.

The key question we face at this point in time is whether or not these recent changes in the industry's research publication output indicate structural and lasting transformations that are reshaping the worldwide corporate science landscape. Or do these trends signal temporary adjustments of business priorities amongst the science-based large firms to cope with cyclic developments affecting the competitive global markets in which they operate? The findings of this exploratory study are obviously suggestive rather than conclusive, and further case studies are necessary to corroborate the tentative conclusions with more detailed information at the firm level and industry level. These first empirical findings do raise a number of important unanswered questions – and related criticism voiced by the scientific community (*Nature*, 2001) – regarding the dynamics of knowledge creation processes in the corporate sector, and interactions with public research organizations in those processes. More specifically, to what extent are the business strategies of firms, and cost projections of in-house basic research, affecting the reservoir of new scientific knowledge and technical knowhow to explore new technological opportunities and generate advanced technologies? How is this process shaping the nature and direction of scientific progress in global science, and the co-evolution of public and corporate science? And have these changes reduced industry's absorptive capacity for new knowledge? Given the wealth of data contained in the still large numbers of papers published by corporate researchers, further empirical research on these topics could certainly benefit from in-depth analyses of industry's contribution to the international scientific and technical journals.

Table 9.1   Selected (parent) companies and country of headquarters – Pharmaceuticals

Abbott Laboratories	USA	IGEN International	USA
Affymetrix	USA	Immunex	USA
Alliance Pharmaceutical	USA	Immunomedics	USA
Amgen	USA	Incyte Genomics	USA
AstraZeneca	UK	Invitrogen	USA
Augustine Medical	USA	Isis Pharmaceuticals	USA
Aventis	USA	Kowa	Japan
Biogen	USA	Kyowa Hakko Kogyo	Japan
BioMerieux	France	Ligand Pharmaceuticals	USA
Bionumerik Pharmaceuticals	USA	Lynx Therapeutics	USA
Biovail	Canada	Merck & Company	USA
Boehringer-Ingelheim	Germany	Millennium Pharmaceuticals	USA
Boots	UK	Neorx	USA
Bristol-Myers Squibb	USA	Neurogen	USA
British Biotech	UK	New England Biolabs	USA
Caliper Technologies	USA	Novartis	Switzerland
Celgene	USA	Novo Nordisk	Denmark
Cell Therapeutics	USA	NPS Pharmaceuticals	USA
Celltech Chiroscience Group	UK	Ono Pharmaceutical	Japan
Cephalon	USA	Pfizer	USA
Chiron	USA	Pharmacia	USA
Chugai Pharmaceutical	Japan	Pharmacopeia	USA
COR Therapeutics	USA	Promega	USA
Corixa	USA	Ribozyme Pharmaceuticals	USA
Corvas International	USA	Roche	Switzerland
Curis	USA	Sanofi-Synthelabo	France
Daiichi Seiyaku	Japan	Schering	Germany
Eisai	Japan	Schering-Plough	USA
Elan	Ireland	Seikagaku	Japan
Eli Lilly	USA	Senju Pharmaceutical	Japan
Emisphere Technologies	USA	Sepracor	USA
Enzon	USA	Shionogi & Company	Japan
Fresenius Chem-Pharm	Germany	Shiseido	Japan
Fujisawa Pharmaceutical	Japan	Sigma-Tau Industrie	Italy
Genentech	USA	Synaptic Pharmaceutical	USA
Genzyme	USA	Taisho Pharmaceutical	Japan
Gilead Sciences	USA	Takeda Chemical	Japan
GlaxoSmithKline	UK	Tanabe Seiyaku	Japan
Guilford Pharmaceuticals	USA	Tularik	USA
Heska	USA	Vertex Pharmaceuticals	USA
Hisamitsu Pharmaceutical	Japan	Wyeth	USA
Human Genome Sciences	USA	Xoma	USA
Hybridon	USA	Yamanouchi Pharmaceutical	Japan
ICOS	USA	Zambon Group	Italy

Sources: DTI's UK 'R&D Scoreboard 2002'; Technology Review/CHI Research's TR
Patent Scorecard 2002.

Table 9.2   Selected (parent) companies and country of headquarters –
Semiconductors

3Com	USA	LSI Logic	USA
Acer	Taiwan	Marconi	UK
Adaptec	USA	Microchip Technology	USA
Altera	USA	Micron Technology	USA
Analog Devices	USA	Mitsubishi Electric	Japan
Apple Computer	USA	Motorola	USA
ATI Technologies	Canada	Murata	
Casio Computer	Japan	Manufacturing	Japan
Cirrus Logic	USA	NEC	Japan
Compaq Computer	USA	Novellus Systems	USA
Conexant Systems	USA	Oce	Netherlands
Dell Computer	USA	Omron	Japan
EMC	USA	Read-Rite	USA
Fujitsu	Japan	Rohm	Japan
Harris	USA	Silicon Graphics	USA
Hewlett-Packard	USA	STMicroelectronics	France
Hitachi	Japan	Storage Technology	USA
Imation	USA	Sun Microsystems	USA
Integrated Device		Taiwan Semiconductor	Taiwan
Technology	USA	Teradyne	USA
Intel	USA	Texas Instruments	USA
Intersil	USA	Tokyo Electron	Japan
Kla-Tencor	USA	Toshiba	Japan
Kyocera	Japan	United	
Lam Research	USA	Microelectronics	Taiwan
Lattice Semiconductor	USA	Western Digital	USA
Lexmark	USA	Xerox	USA
Linear Technology	USA	Xilinx	USA

Sources: DTI's UK 'R&D Scoreboard 2002'; Technology Review/CHI Research's TR
Patent Scorecard 2000.

NOTES

1.   Bert van de Wurff and Erik van Wijk are gratefully acknowledged for their contributions
to the CWTS's database of corporate research papers. Thed van Leeuwen was of great
help in conducting the data analyses.
2.   Industrial basic research, is loosely defined as research not related to current corporate
products, and covers both longer term scientific research and engineering research. This
kind of research is often driven by a strategic vision of the market with a three to five
year time horizon. In certain fast moving technology areas, the terms 'research' and
'long term' are no longer coupled in the traditional way; the commitment to research

is long term, but the research projects and programmes themselves may have short term objectives and deliverables.

3. A company's knowledge base is comprised of the accumulated sum of knowledge on which the advance of the firm relies; – includes not only codified knowledge, but also tacit knowledge and knowledge embedded in equipment, instruments and the plant. The former refers to knowledge that has been reduced to a written and transmittable form, while the latter refers to knowledge that exists subconsciously in the human mind, and is acquired through experience, imitation, and observation, and can be transferred only by personal contact (David and Foray, 1995; Nonaka and Takeuchi, 1995).

4. The fraction of business funding in the OECD countries of research conducted in the university sector has increased from 1.4 per cent of the total business R&D funding to 1.7 per cent during the years 1995–99 (OECD, 2000).

5. The rule of 'scientific priority' in the scientific communication process identifies the prime knowledge producer and the moment of publication and builds a reputation, which is crucial for obtaining recognition in the scientific community, receiving tenure, entering networks and receiving grants. Granting researchers authorships and associated 'moral' intellectual property rights of their fruits of labour, rather than granting them exclusive intellectual rights to the knowledge, results in the 'knowledge market dilemma', where researchers need to be efficient and productive in their research efforts while having little or no chance of keeping the financial rewards for themselves. It enables the creation of a private asset for the 'discoverer' resulting from the very fact of giving up exclusive rights. The need to be identified and recognized as the discoverer impels speedy and full disclosure (Dasgupta and David, 1994). This reward mechanism creates races or competitions, still involving full release of the knowledge. Full disclosure also acts as a quality control system since publicly published results can be duplicated and checked by other scientists. The rule of priority combined with the open science system guarantees dissemination without reducing motivation, while improving the quality of research and cumulative and collective scientific advance.

6. Similar reasons may exist for companies to reveal (research-based) innovation-related information (see e.g. Harhoff et al., 2003).

7. For example, an industry/university research paper written by five researchers: two from different Pfizer labs, one from a Pharmacia lab, and two from different research groups at Cambridge University, will increment the count once for Pfizer, once for Pharmacia, and once for Cambridge.

8. OECD background papers contain provisional data pointing towards decreases in the USA and Japan for corporate expenditure for basic research (OECD, 2001).

9. Corporate R&D expenditure by US-based companies increased by some 10 per cent each year in the second half of the 1990s. Data collected by the US National Science Foundation show that basic research accounted for 9 per cent of the total corporate R&D spending in 2000, shorter term ('applied') research for 20 per cent, and the remainder of 71 per cent for technical development (NSB, 2002). From 1995 to 2000, aggregate R&D expenditure increased by 63 per cent, but basic research rose by 142 per cent (Larson, 2001). The exceptionally strong growth rate is claimed to be a reflection of increased funds for corporate profits and available cash for future investments and risk taking, decreasing product cycle times and strong competition in increasingly global markets.

10. The exponential growth of patent citations in research papers in international scientific and technical journals in the years 1996–98 is in part due to changes in the US patent law in 1995 (NSB, 2002, pp. 5–53).

11. Although these corporate co-publications do not list authors from the public sector, the research efforts reported in these papers may well include significant contributions from universities or research institutes. These sources may turn up in the acknowledgements or are 'hidden' in the list of references.

12. The set of interfirm co-publications includes co-authored papers listing different national affiliates or subsidiaries of the same (ultimate) parent company located in different countries.

13. The industry's influence on the scientific progress in the global research system, and its contribution to the open scientific publication system, may therefore in part be hidden from public scrutiny. Moreover, a firm's research partners at universities and public sector research institutes might still be publishing results of joint efforts in international journals without mentioning the (monetary) resources supplied by the corporate sector. For this reason, several high profile scientific journals, like *Nature*, have implemented editorial policies forcing authors to explicitly acknowledge such ties with industry.

REFERENCES

Arrow, K. (1962), 'Economics of welfare and the allocation of resources for invention', in National Bureau of Economic Research, *The Rate and Direction of Inventive Activity*, Princeton, NJ: Princeton University Press.

Arundel, A. and R. Garrelfs (eds) (1997), *Innovation Measurement and Policies*, EIMS publication 50, Brussels: European Commission.

Barney, J. (1991), 'Firm resources and competitive advantage', *Journal of Management*, **17** (1), 99–120.

Beise, M. and H. Stahl (1999), 'Public research and industrial innovations in Germany', Research Policy, **28** (4), 397–422.

Buderi, R. (2000a), *Engines of Tomorrow: How the World's Best Companies are Using their Research Labs to Win the Future*, New York: Simon and Schuster.

Buderi, R. (2000b), 'Funding central research', *Research-Technology Management*, **43** (4), 18–25.

Cockburn, I. and R. Henderson (1998), 'Absorptive capacity, co-authoring behaviour, and the organisation of research in drug discovery', *Journal of Industrial Economics*, **46** (2), 157–82.

Cohen, W. and D. Levinthal (1989), 'Innovation and learning: the two faces of R&D', *The Economic Journal*, **99** (397), 569–96.

Cohen, W. and D. Levinthal (1990), 'Absorptive capacity: a new perspective on learning and innovation', *Administrative Science Quarterly*, **35** (1), 128–52.

Coombs, R. and L. Georghiou (2002), 'A new "industrial ecology"', *Science*, **296** (5567), 471.

Dasgupta, P. and P. David (1994), 'Towards a new economics of science', *Research Policy*, **23** (5), 487–521.

David, P. and D. Foray (1995), 'Accessing and expanding the science and technology knowledge base', *OECD STI Review*, **16**, 13–68.

Geullec, D. (2001), 'Basic research: statistical issues', OECD Document DSTP/EAS/STP/NESTI 38, Paris: OECD.

Gibbons, M., C. Limoges, H. Nowotony, S. Schwartzman, P. Scott and M. Trow (1994), *The New Production of Knowledge: The Dynamics of Science and Research in Contemporary Societies*, London: Sage.

Godin, B. (1996), 'Research and the practice of publication in industries', *Research Policy*, **25** (4), 587–606.

Grant, R. (1995), *Contemporary Strategy Analysis: Concepts, Techniques, Applications*, Cambridge, MA: Blackwell Press.

Griliches, Z. (1979), 'Issues in assessing the contribution of research and development to productivity growth', *The Bell Journal of Economics*, **10** (1), 92–116.

Griliches, Z. (1986), 'Productivity, R&D and basic research at the firm level in the 1970s', *American Economic Review*, **76** (1), 141–154.

Hagedoorn, J., A. Link and N. Vonortas (2000), 'Research partnerships', *Research Policy*, **29** (4/5), 567–86.

Hall, B. (1996), 'The private and social returns to research and development', in B. Smith and C. Barfield (eds), *Technology, R&D and the Economy*, Washington, DC: Brookings Institution.

Halperin, M. R. and A. K. Chakrabarti (1987), 'Firm and industry characteristics influencing publications of scientists in large American companies', *R&D Management*, **17**, 167–73.

Harhoff, D., J. Henkel and E. von Hippel (2003), 'Profiting from voluntary information spillovers: how users benefit freely by revealing their innovations', *Research Policy*, **32** (10), 1753–69.

Hicks, D., T. Ishizuka, P. Keen and S. Sweet (1994), 'Japanese corporations, scientific research and globalization', *Research Policy*, **23** (4), 375–84.

Hicks, D. and J. Katz (1997), 'The changing shape of British industrial research', SPRU/STEEP special report no. 6, University of Sussex.

Jaffe, A. (1989), 'Real effects of academic research', *American Economic Review*, **79** (5), 957–70.

Katz, J. S. and B. R. Martin (1997), 'What is research collaboration?' *Research Policy*, **26** (1), 1–18.

Larson, C. (2001), 'R&D and innovation in industry', in *Research and Development FY 2002*, American Association for the Advancement of Science report XXVI, Washington, DC.

Levin, R. C., A. K. Klevorick, R. R. Nelson and S. G. Winter (1987), 'Appropriating the returns from industrial research and development', *Brookings Papers on Economics Activity*, **3**, 783–820.

Lim, K. (2001), 'The relationship between research and innovation in the semiconductor and pharmaceuticals industries (1981–1997)', National University of Singapore working paper.

Mansfield, E. (1981), 'Composition of R&D expenditures: relationship to size of firm, concentration and innovative output', *Review of Economics and Statistics*, **63** (4), 610–15.

Mansfield, E. (1991), 'Academic research and innovation', *Research Policy*, **20** (1), 1–12.

Meyer-Krahmer, F. and U. Schmoch (1998), 'Science-based technologies: university–industry interactions in four fields', *Research Policy*, **27** (8), 835–51.

Mowery, D. C. (1983), 'Economic theory and government technology policy', *Policy Sciences*, **16**, 29–43.

National Science Board (NSB) (2002), *Science and Engineering Indicators 2002*, (NSB-02-01), Arlington, VA: National Science Foundation.

Nature (2001), 'Is the university–industrial complex out of control?', *Nature*, **409** (6817), 119.

Nelson, R. R. (1959), 'The simple economics of basic scientific research', *Journal of Political Economy*, **67**, 297–306.

Nelson, R. R. (1989), 'What is private and what is public about technology?', *Science, Technology and Human Values*, **14**, 229–41.

Nelson, R. R. (1990), 'Capitalism as an engine of progress', *Research Policy*, **19**, 193–214.

Nonaka, I. and H. Takeuchi (1995), *The Knowledge-creating Company: How*

Japanese Companies Create the Dynamics of Innovation, New York: Oxford University Press.

OECD (2000), *Main Science and Technology Indicators (MSTI) Database*, November, Paris: OECD.

OECD (2001), 'Changing business strategies for R&D and their implications for science and technology policy: OECD background paper and issues paper, OECD document DSTP/STP (2001) 29, Paris: OECD.

Pavitt, K. (1991), 'What makes basic research economically useful?', *Research Policy*, **20** (2), 109–19.

Rosenberg, N. (1990), 'Why do firms do basic research with their own money?', *Research Policy*, **19** (2), 165–74.

Salter, A. J. and B. R. Martin (2001), 'The economic benefits of publicly funded basic research: a critical review', *Research Policy*, **30** (3), 509–32.

Small, H. and E. Greenlee (1977), *A Citation and Publication Analysis of US Industrial Organisations*, Washington, DC: National Science Foundation.

Teece, D. J. (1987), 'Profiting from technological innovation: implications for integration, collaboration, licencing and public policy', in D. J. Teece (ed.), *The Competitive Challenge*, Cambridge, MA: Ballinger.

Tijssen, R. J. W. (1998), 'Quantitative assessment of large heterogeneous R&D networks: the case of process engineering in the Netherlands', *Research Policy*, **26** (X), 791–809.

Tijssen, R. J. W. (2002), 'Science dependence of technologies: evidence of inventions and their inventors', *Research Policy*, **31** (4), 509–26.

Tijssen, R. J. W., Th. N. Van Leeuwen and J. C. Korevaar (1996), 'Scientific publication activity of industry in the Netherlands', *Research Evaluation*, **6**, 1–15.

Varma, R. (2000), 'Changing research cultures in US industry', *Science, Technology and Human Values*, **25** (4), 395–416.

Zucker, L., M. Darby and M. Brewer (1998), 'Intellectual human capital and the birth of US biotechnology enterprises', *American Economic Review*, **88** (1), 290–306.

PART THREE

Long-term technological change
and the economy

10. Making (Kondratiev) waves: simulating long-run technical change for an Integrated Assessment system[1]

Jonathan A. Köhler

1. INTRODUCTION

A world macroeconomic model is being developed to investigate policies for climate change and sustainable development. To analyse climate change policy, a timescale of 100 years is necessary, because changes in CO_2 concentrations, which are now strongly influencing the atmosphere, become significant over a time period of 50–100 years or more.

This raises particular difficulties for economic modelling. Looking back over the last 200 years, the socioeconomic system seems to be characterized by ongoing fundamental change, rather than convergence to an equilibrium state. Our opinion is that over such a long time period, a neoclassical economic model incorporating a long-term equilibrium for the world economy is inappropriate. It is necessary instead to consider the dynamic processes of socioeconomic development. These processes have been called 'Kondratiev waves' in the literature on long-term economic development (Freeman and Louçã, 2001).

This chapter suggests a quantitative theory of long-term technical change. It will be part of a global macroeconometric model. Dewick et al. (2004) describe the process of assessing the future technologies to which this theory will be applied. In Section 2, the case is made for a disequilibrium analysis, in the spirit of the evolutionary economists. In Section 3, a (descriptive) theory of long-term economic change is discussed and an interpretation suitable for incorporation in a macroeconomic modelling framework introduced. A simple model of a growth sector is introduced in Section 4 and some preliminary results are given in Section 5. Section 6 concludes and points towards further developments.

2.   A THEORY OF INDUSTRIAL REVOLUTIONS

The requirement for a macroeconomic model to 2100 leads us to the conclusion that there is a need for a detailed analysis of the macroeconomics of long-term changes. This is in sharp contrast to current economic models (and the economic components of Integrated Assessment Models) used for climate change policy analysis. Typical examples are the DICE economic model (Nordhaus, 1994) and the IMAGE Integrated Assessment Model (Alcamo et al., 1998), but an extensive review of this literature is beyond the scope of this chapter. A review of the models and their features is Barker et al. (2002).

Our central argument is that, since 1750, socioeconomic activity has been characterized by a series of fundamental changes in technology, institutions and society. This follows the earlier thinking of Kondratiev, Schumpeter and more recently evolutionary economists (Arthur, 1994; Day, 1994; Dosi, 2000; Freeman and Soete, 1997; Nelson and Winter, 1982; Perez, 1983; Silverberg and Soete, 1994) and economic historians (David, 1993).

Freeman and Louçã (2001) include a history of economic thought in this area, starting from a critique of cliometrics, the use of econometric methods in economic historical analysis. They cover the ideas of Kondratiev and Schumpeter in particular, who were the leading early figures in economic analysis of long-term economic changes. Kondratiev formulated the hypothesis that there were long waves in capitalist development, now called 'Kondratiev waves'. He undertook one of the first quantified statistical analyses of long-term economic data and identified an approximate dating of the long term upswings and downswings with distinctive characteristics in capitalist economies. Schumpeter applied economic theoretical ideas to the study of long-term economic change, in a search for an economic theory of the processes of economic change in economic history.

The current (numerical) models of long-term technical change have often been developed in the tradition of evolutionary economics, often using the mathematics developed for dynamic processes in biology. For example, Arthur (1994) applied a random process to the cost reduction in a competition between two technologies to demonstrate that one technology would eventually dominate the market with 100 per cent probability and this would not necessarily be the most effective technology, (the phenomenon of 'lock-in').

The problem with the models in this field is that they are theoretical and conceptual, rather than dependent upon empirical analysis. They are not based on the assumption of economic rationality or a Walrasian economic structure, so there is no consensus about what a reasonable theoretical structure might be. These features mean that such models cannot be easily parameterized and estimated/calibrated for inclusion in a numerical macroeconomic model that tries to deliver 'real world' policy conclusions. Also, this field has concentrated on industrial structure, studying competition between firms, often with different technologies. This is vital for an understanding of the processes of change, but means that the models are microeconomic models of competition within a sector of the economy and not macroeconomic. There has been very little work in this tradition on macroeconomic models; (with the exception of Boyer, see for example Boyer, 1988).

Empirical analysis in the normal sense (for economists) of econometrics has some serious limitations for this analysis. Indeed, the econometric approach initiated by Kondratiev is specifically rejected by Freeman and Louçã (2001). Econometric models depend on looking backwards to develop the model and then can only extrapolate from past trends into the future. The data for long-term economic change are necessarily sketchy and econometric methods are therefore ill suited to such broad analyses, particularly when a view of the long-term future is needed and fundamental changes in the socioeconomic system are postulated.

To summarize, there is no suitable and generally accepted theory of long-term technical change for incorporation in a macroeconomic modeling structure. However, there is now a good descriptive theory, which is intended to provide an economic history perspective of long-term change: Freeman and Louçã (2001). They argue that Kondratiev waves involve a process of dynamic interaction between five subsystems: science, technology, economy, politics and culture. For our purpose of developing a quantitative model, it is only realistic to try and model technology and economy. The impacts and feedbacks through the other subsystems will be reflected qualitatively in the macroeconomic model structure and through scenarios. The objective of our model is to interpret this descriptive theory in quantitative terms, as far as is plausible, in the context of the macroeconomic analysis outlined in the introduction.

3. FREEMAN AND LOUÇÃ'S THEORY OF LONG-TERM CHANGE

Freeman and Louçã (2001) identify five waves of technology and socioeconomic activity since the industrial revolution in the UK:

1. Water-powered mechanization of industry;
2. Steam-powered mechanization of industry and transport, based on iron and coal;

3. Electrification of industry, transport and the home, with steel as a core input;
4. Motorization of transport, civil and war economies, with industrial chemicals and oil as core inputs;
5. Computerization of the economy.

Following Perez (1983), they characterize Kondratiev waves as a succession of new technology systems (Freeman and Louçã, 2001, pp. 147–8):

1. For each long wave, there are 'core inputs' such as iron for the railway wave, that become very cheap and universally available. This opens up new possibilities of production factor combinations. The sector producing these inputs is the 'motive branch'.
2. New products based on the new factor combinations give rise to new industries whose growth drives the whole economy, for example railways and the associated production of rails, locomotives and railway equipment.
3. There are new forms of organization of production brought about by the new industries and products, a new 'technoeconomic paradigm'.
4. Such a fundamental change will lead to a period of turbulent adjustment from the old paradigm to the new.

Freeman and Louçã (2001) identify the following six phases in the life cycle of a technology system:

1. Laboratory/invention;
2. Decisive demonstration(s) of technical and commercial feasibility. Continuing with the railways example, the opening of the Liverpool and Manchester railway in the UK in 1830 is an outstanding example;
3. Explosive, turbulent growth, characterized by heavy investment and many business start-ups and failures. There is a period of structural crisis in the economy as society changes to the new organizational methods, employment and skills and regime of regulation, brought about in response to the new technology;
4. Continued high growth, as the new technology system becomes the defining characteristic of the economy;
5. Slowdown, as the technology is challenged by new technologies, leading to the next crisis of structural adjustment;
6. Maturity, leading to a (smaller) continuing role of the technology in the economy or slow disappearance.

As can be seen from Table 10.1 and illustrated in Figure 10.1, phases two to five have been found to take roughly 50 years. In phase one, which is of

Table 10.1 Condensed summary of Kondratiev waves

Wave	Decisive innovations	Carrier branches	Core input(s)	Infrastructure	Management; organization	Upswing (boom)	Downswing (crisis of adjustment)
1. Water-powered mechanization of industry	Arkwright's mill 1771	Cotton spinning Iron	Iron, Cotton Coal	Canals Turnpike roads Sailing ships	Factory systems Entrepreneurs Partnerships	1780s–1815	1815–1848
2. Steam-powered mechanization of industry and transport	Liverpool and Manchester railway 1830	Railways Steam engines Machine tools Alkali industry	Iron Coal	Railways Telegraph Steamships	Joint stock companies Subcontracting to craft workers	1848–1873	1873–1895
3. Electrification of industry, transport and the home	Bessemer steel process 1875; Edison's electric power plant 1882	Electrical equipment Heavy engineering Chemicals Steel products	Steel Copper Metal alloys	Steel railways Steel ships Telephone	Specialized, professional management systems; 'Taylorism'; Giant firms	1895–1918	1918–1940
4. Motorization	Ford's assembly line 1914; Burton process for cracking oil 1913	Cars Aircraft Internal combustion engines Oil refining	Oil Gas Synthetic materials	Radio Motorways Airports Airlines	Mass production and consumption 'Fordism' Hierarchies	1941–1973	1973–?
5. Computerization of the economy	IBM computers 1960s Intel processor 1972	Computers Software Telecommunications equipment	Silicon 'chips' (integrated circuits)	Internet	Networks: internal, local, global	approx. 1980–?	

Source: Freeman and Louçã (2001, p. 141).

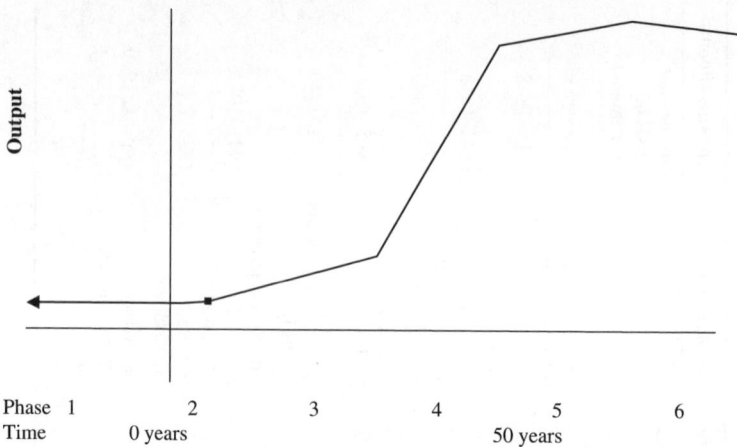

Figure 10.1   Phases in the life cycle of a technology system

indeterminate length, there is a negligible macroeconomic effect. The timing of the invention leading to a breakthrough in the technology and the application in a 'decisive demonstration' is more or less random, viewed from a economic perspective. It is phases two to five that lead to the Kondratiev waves.

This view of Kondratiev waves leads Freeman and Louçã (2001) to the following conclusions/hypotheses:

1.   There is a period in which there are technological and/or organizational innovations offering very high profits in a period of general decline in the rate of profit (phases two and three).
2.   There are recurring structural crises of adjustment, structural unemployment, social unrest as society switches from one technology system to the next (phases three and five).
3.   The new technological system is associated with a change of regulatory and institutional regime.
4.   Each wave generates a new cohort of very large firms, compared to the industrial organization of the previous wave, in the new sector(s).
5.   There is a high level of industrial unrest in two phases:
    phase three: structural adjustment, with a mismatch of skills, as workers in 'old industries' are made redundant while new skills are often only acquired by new entrants to the workforce;
    phase five: decline in rate of profit with strong unions.

4. A SIMULATION MODEL OF A SECTOR WITH A TECHNOLOGICAL REVOLUTION

The most difficult challenge in interpreting this descriptive theory of Kondratiev waves is the very large extent to which each wave has unique features of organization and sectoral activity, as can be seen in Table 10.1. This problem has been addressed using the following approach.

The necessary features of the technology model are:

1. It should generate the output path over time in the six phases.
2. Following Criqui et al. (1999) on how to model endogenous technical change, it should incorporate or at least take into consideration exogenous inventions, supply (R&D and technological opportunities) and demand (new products and markets) inducement factors. It should model path dependency (learning by doing and increasing returns).
3. It should have declining production costs in the new sectors, incorporating endogenous technical change through R&D expenditure, investment and learning by doing, that is, investment impacts (following Grübler et al., 1999).

The key assumptions of the theory are:

1. The new technology is taken up by a 'carrier branch' of industry, to use Perez's 1983) terminology.
2. The new technology is embodied in a 'core input', whose price suddenly drops dramatically.
3. This leads to 'super normal' profits in the carrier branch, which then leads to an expectation of high profits, resulting in many start-ups of firms with high R&D expenditure and investment.
4. Demand is a function of market size and relative prices.
5. R&D and investment are a function of expected profits.

4.1 Theory

4.1.1 Production function
The theory will form part of a dynamic macroeconomic model and must therefore link up with macroeconomic variables. The simulation is of a sub-branch of an industrial sector in which there is a technological breakthrough, leading to a sudden drop in production costs. Following this, a 'learning by doing' cost function means that production costs fall with cumulated investment and R&D expenditure, following the results of IIASA (Grübler et al., 1999).

Production cost (at time t)

$$totcurr_t = totcurr_{t-1} + rdcurr + inv_{t-1} \tag{1a}$$
$$cost_t = cost1 * \exp(-1.0*cost2*totcurr_t) + cost3 \tag{1b}$$

where:

totcurr	is total knowledge capital
rdcurr	is current R&D spending
inv	is investment
cost	is production cost; *cost*1, *cost*2, *cost*3 are parameters

4.1.2 Supply function

Supply exhibits a lagged response with an adaptive expectations formulation. The lagged response of supply reflects the rigidities in industrial behaviour. An example is given in Cox and Popken (2002) where they state that the telecoms industry plans production 6–18 months in advance. It depends on the profitability as indicated by the mark-up factor in the previous period $t-1$ and is bounded upwards by the current capital stock.

$$supp_t = s3*(s1*k_t*(1-\exp(-1.0*s2*m))-supp_{t-1})+supp_{t-1} \tag{2}$$

where:

supp	is supply; *s*1, *s*2 and *s*3 are parameters
k	is capital stock (see equation 4)
m	is the markup pricing factor (see equation 6)

Note one unusual feature: supply is dependent on mark-up rather than sales price. As this is not an equilibrium, the price does not give a direct signal of whether goods are sold at a profit or loss. This signal is given by the mark-up factor. The underlying argument is the same as in a conventional model, firms sell as much as they can until (marginal) profits become zero.

4.1.3 GDP growth (mainly exogenous)

$$gdp_t = gdp_{t-1}*(1+gdp1) + gdp2*supp_t \tag{3}$$

where:

gdp	is domestic output
*gdp*1	is an exogenous growth parameter
*gdp*2	is a factor allowing for the macroeconomic impact of the technology sector

4.1.4 Capital accumulation

$$k_t = k_{t-1}*(1-dep)+inv_{t-1} \tag{4}$$

where:
- k is capital stock
- *dep* is the depreciation of capital per period
- *inv* is investment in this new technology (see equation 7)

4.1.5 Demand and price determination (simultaneous)

Given the previously determined supply, price is found using a mark-up pricing rule. Demand in the current version of the model is a linear decreasing function of price. Price and demand are determined simultaneously, given supply and current production cost. Note also that for mathematical convenience, the markup m differs from the conventional markup. The conventional markup is expressed as a percentage addition to the cost (price = cost + %markup). In this model, the markup is a multiplying factor of cost (price=cost*markup factor).

$$dem_t = (gdp_t*d1)-d2*cost_t*m_t \tag{5}$$
$$m_t = m1*dem_t/supp_t \tag{6}$$

where:
- *dem* is demand

4.1.6 Investment function

Finally investment is a nonlinear function – a quadratic – of profitability.

$$pi_t = (m_t-1)*cost_t*dem_t \tag{7a}$$
$$inv_t = (1/(1+r_t))*inv2*pi_t*abs(pi_t) \tag{7b}$$

where:
- *pi* is profit
- *r* is interest rate
- *inv2* is a constant parameter

Note that if profit is negative, the inclusion of abs(pi) means that investment is also negative. This can be interpreted as an expression of the fact that share prices may fall as well as rise, impacting on the ability of firms to purchase capital goods.

5.   PRELIMINARY RESULTS

A selection of initial results are shown in Figures 10.2–10.5. Details of the parameterization are available from the author. The horizontal axes can be thought of as years, the vertical axes are in real prices. The figures plot investment, supply and demand over time. Figure 10.2 demonstrates that the model is capable of generating investment bubbles and an initial boom – or rapid expansion of capacity. It also shows a fluctuating expansion of activity in the long term. In this parameterization, long-term growth is determined by the positive feedback of increases in supply increasing GDP, which shifts the demand function upwards. The main features of the first four of the stages of a Kondratiev wave are therefore shown. The slowing down due to market saturation and increasing competitiveness in the long term will require a more realistic modelling of demand allowing for market saturation. This is discussed further in the conclusions.

The model is capable of generating unstable behaviour and the chaotic properties associated with nonlinear dynamic systems. Figures 10.3–10.5 demonstrate some of the range of behaviours that can be represented, even by such a simple economic model. Figure 10.3 shows a cyclical growth path with an eventual collapse of the price. Figure 10.4 illustrates a case in which the industry fails; the initial cost reduction is not great enough for demand to take off. Finally, Figure 10.5 shows a case closely related to Figure 10.4. After an initial decline, the cost and price reductions following a continued relatively low rate of investment are just sufficient to spark a demand growth.

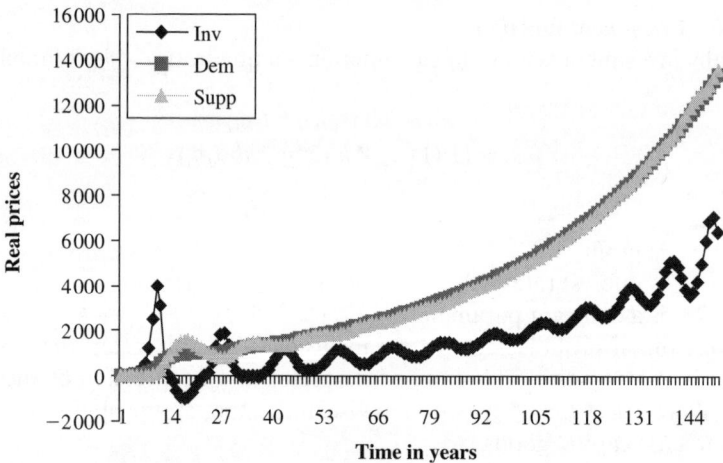

Figure 10.2   Investment, supply and demand over time

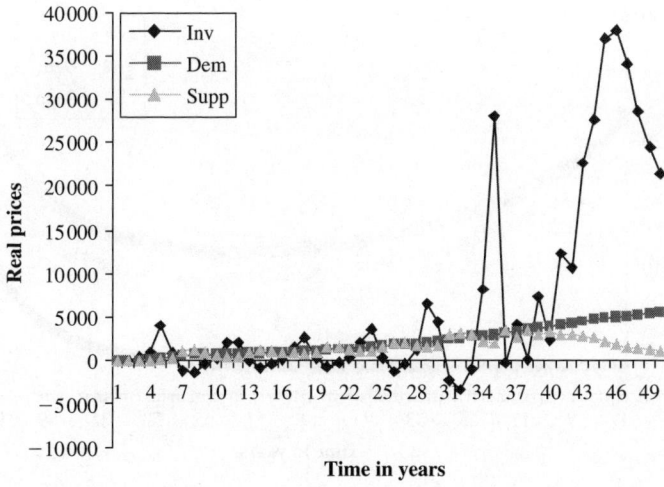

Figure 10.3   Cyclical growth path with an eventual collapse of price

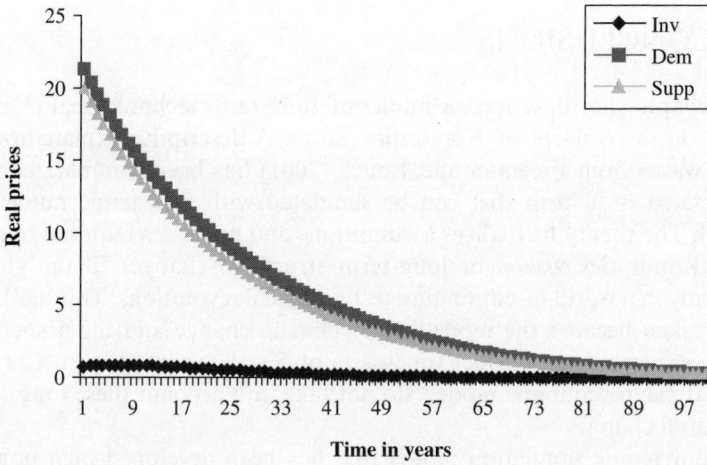

Figure 10.4   Industry failure

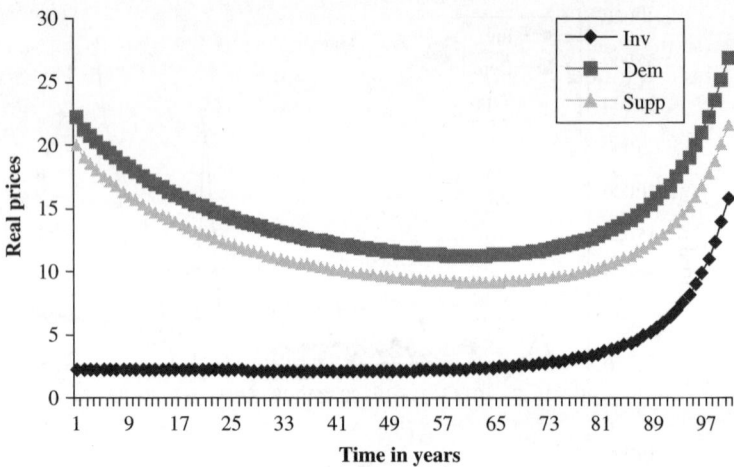

Figure 10.5   Initial decline in cost and price following a relatively low rate of investment are sufficient to spark demand growth

6.   CONCLUSIONS

This chapter has described a model of long term technological change, based on the concept of Kondratiev waves. A descriptive explanation of these waves from Freeman and Louçã (2001) has been summarized and interpreted in a form that can be simulated with a dynamic numerical model. The theory formalizes assumptions and processes required to generate Kondratiev waves, or long-term structural changes to the global economy in a world of continuing technological revolutions. This has been undertaken because the modelling of climate change and the associated policy issues has to consider timescales of 50–100 years at least. Current general macroeconomic models do not take into account these long-term structural changes.

The dynamic simulation model that has been developed incorporates some unusual features, which enable it to generate a wide variety of development paths of an industry. It generates the boom phase of a Kondratiev wave together with the investment bubbles that accompany the early phases in a new wave. Its results are dependent on increasing returns to scale in production costs, a lagged response to supply to the market situation and a rapid response of investment to profitability. The main shortcoming of this model is the linear demand response. Particularly in the long term, markets become more competitive as they become saturated.

Therefore, a dynamic demand model allowing for a slowing down of the increase in demand may deliver new insights into the growth behaviour. Finally, the model must be calibrated against historical data on earlier Kondratiev waves before it can be used as input into a long-term view of economic change.

NOTE

1. This work is funded under the UK Tyndall Centre research theme 'Integrating Frameworks'.

REFERENCES

Alcamo, J., R. Leemans and E. Kreileman (eds) (1998), *Global Change Scenarios of the 21st Century: Results from the IMAGE 2.1 Model*, London: Elsevier Science.

Arthur, W. B. (1994), *Increasing Returns and Path Dependence in the Economy*, Ann Arbor, MI: University of Michigan Press.

Barker, T., J. Koehler and M. Villena (2002), 'The costs of greenhouse gas abatement: a meta-analysis of post-SRES mitigation scenarios', *Environmental Economics and Policy Studies*, **5** (2), 135–66.

Boyer, R. (1998), 'Technical change and the theory of Regulation', in G. Dosi, C. Freeman, R. Nelson, G. Silverberg and L. Soete (eds), *Technical Change and Economic Theory*, London: Pinter, pp. 67–94.

Cox, L.A. and D.A. Popken (2002), 'A hybrid system-identification method for forecasting telecommunications product demands', *International Journal of Forecasting*, **18** (4), 647–71.

Criqui, P., N. Kouvaritakis, A. Soria and F. Isoard (1999), 'Technical change and CO_2 emission reduction strategies: from exogenous to endogenous technology in the POLES model', in P. Criqui (ed), *Le progrès technique face aux défis énergétiques du futur*, Paris: Colloque européen de l'énergie de l'AEE, pp. 473–88.

David, P. A. (1993), 'Path-dependence and predictability in dynamic systems with local network externalities: a paradigm for historical economics', in D. Foray and C. Freeman (eds), *Technology and the Wealth of Nations: The Dynamics of Constructed Advantage*, London: Pinter, pp. 208–31.

Day, R. H. (1994), *Complex Economic Dynamics*, Cambridge, MA and London: MIT Press.

Dewick, P., K. Green and M. Miozzo (2004), 'Technological change, industrial structure and the environment', *Futures*, **36** (3) (March), 267–93.

Dosi, G. (2000), *Innovation, Organization and Economic Dynamics: Selected Essays*, Cheltenham, UK and Northampton, MA: Edward Elgar.

Freeman, C. and F. Louçã (2001), *As Time Goes By*, Oxford: Oxford University Press.

Freeman, C. and L. Soete (1997), *The Economics of Industrial Innovation*, 3rd edn, London: Pinter.

Grübler, A., N. Nakicenovic and D. G. Victor (1999), 'Dynamics of energy technologies and global change', *Energy Policy*, **27** (5), 247–80.

Nelson, R. R. and S. G. Winter (1982), *An Evolutionary Theory of Economic Change*, Cambridge, MA: Harvard University Press.

Nordhaus, W. (1994), *Managing the Global Commons: The Economics of Climate Change*, Cambridge, MA: MIT Press.

Perez, C. (1983), 'Structural change and the assimilation of new technologies in the economic and social system', *Futures*, **15** (5), 357–75.

Silverberg, G. and L. Soete (eds) (1994), *The Economics of Growth and Technical Change: Technologies, Nations, Agents*, Aldershot, UK and Brookfield, USA: Edward Elgar.

11. Nonlinear dynamism of innovation and knowledge transfer

Masaaki Hirooka[1]

1. INTRODUCTION

This chapter proposes a new concept for innovation and knowledge transfer. This approach offers a powerful tool to analyse ongoing innovation and knowledge transfer in a rapidly changing global economy.

In the economic study of innovation so far, the diffusion of innovation, market trends and the behaviour of firms have been intensively discussed. There is, however, a long latent period of technology development before the beginning of the diffusion of innovation. This technology development period has not been sufficiently treated: it is a black box. This chapter throws light on this technology development period and thus it becomes possible to discuss an innovation paradigm as a comprehensive system consisting of two periods of technology development and product diffusion.

One of the important findings of this study is the nonlinear nature of innovation and knowledge transfer. The market for innovation products reaches an ultimate maturity which never exceeds some limit. This relationship is well described by a logistic equation and we designate this locus described by a logistic equation as a 'trajectory'. A new finding presented in this chapter is that technology development itself has a nonlinear nature and can be described by a logistic equation. This is the main subject of this chapter; and central to this is the knowledge transfer phenomenon in the course of innovation.

This chapter is organized as follows. As a background, section 1 introduces the concept of innovation diffusion as a logistics curve and offers evidence of the diffusion coefficient of 17 products. Section 2 examines if the logistic relationship holds for the technology (development) trajectory and the (product) development trajectory: these two stages precede the diffusion trajectory. Section 3 describes the three trajectories (collectively referred to as an innovation paradigm) for electronics, biotechnology and synthetic dyestuffs. Section 4 discusses the development trajectory in more detail, focusing on the role of universities, venture business and national systems of innovation with respect to electronics, biotechnology and synthetic dyestuffs industries. Section 5 explains the implications of the nonlinearity findings for innovation studies and section 6 presents some concluding remarks.

2.   LOGISTIC DYNAMISM OF INNOVATION DIFFUSION

The economics of technological change has been discussed for a long time since Schumpeter pointed out the importance of technological innovation for economic development. Schumpeter (1939) ascribed the formation of Kondratiev's long waves to technological innovation in his book "Business Cycles". Since the Industrial Revolution, the economy has actually developed by various innovations which build economic infrastructures. The diffusion of innovation to make a market is described by a logistic equation as first pointed out by Griliches (1957) and many economists have confirmed this relationship, (for example Fisher and Pry, 1971; Mansfield, 1961, 1963, 1968; Marchetti, 1979, 1980, 1988; Marchetti et al., 1995, 1996; Marchetti, 2002; Metcalfe, 1970; Modis, 1992; Nakicenovic and Grübler, 1991). Some authors, such as David (1975), Davies (1979), Metcalfe (1981, 1994), and Stoneman (1983) proposed modified models, to, for example, explain the correlation between demand and supply in the economy.

Hirooka and Hagiwara (1992) extensively studied the diffusion of various innovation products by expressing the diffusion phenomenon as a logistic equation. This chapter begins by briefly discussing the results of these analyses for the diffusion of innovation.

2.1   Logistic Equation

The logistic equation for product diffusion is expressed by the formula (1):

$$dy/dt = a\, y\, (y_0 - y) \tag{1}$$

where y is product demand at time t,
       y_0 is the ultimate market size,
and a is a constant
The solution of this nonlinear differential equation is (2):

$$y = y_0 / [1 + C \exp(-ay_0 t)] \tag{2}$$

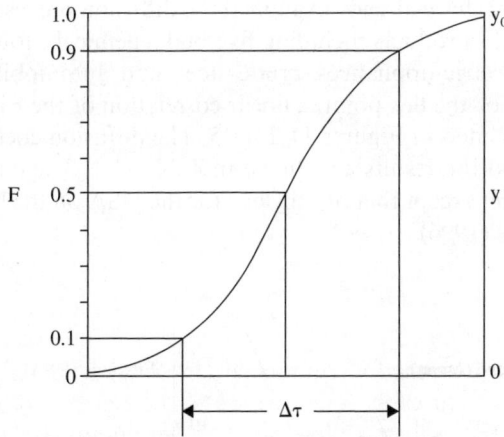

Figure 11.1 Logistic description of innovation and time span Δτ

If the logistic equation is expressed by the fraction $F = y/y_0$, the equations (1), (2) are represented by the formulae (3), (4):

$$dF/dt = \alpha F (1-F) \qquad (3)$$
$$F = 1 / [\, 1 + C \exp (-\alpha t)] \qquad (4)$$

This equation was transformed by Fischer and Pry (1971) to make a linear relation on time t which is formulated by equation (5):

$$\ln F / (1-F) = \alpha t - b \qquad (5)$$

The ultimate market size y_0 is determined by the flex point of the logistic curve, $y_0 / 2$, which is the secondary differential coefficient of (1), and the adaptability of the logistic equation is examined by the linearity of the Fisher–Pry plot. The α is the diffusion coefficient of the product to the market. If the time span between $F = 0.1$ and $F = 0.9$, is conveniently taken to express the spread of the logistic curve, this is a conventional expression of the time dependence of the product diffusion to the market as shown in Figure 11.1. This kind of treatment was also used by Marchetti (1979, 1988).

2.2 Logistic Dynamism of Product Diffusion

Before discussing the period of technology development, it is important to describe a new concept of the diffusion process of innovation. Hirooka and

Hagiwara (1992) have already examined the diffusion process by the above procedure for 17 products including five bulk chemicals, four engineering plastics, six electric appliances, crude steel, and automobiles. From the determination of the flex point, a linear correlation of the Fisher–Pry plot is examined as cited in Figures 11.2–11.5. The diffusion coefficient, α was determined and the results are shown in Table 11.1. The data are for the Japanese market except that of ethylene for the USA on the basis of MITI (2000) and UN (1996).

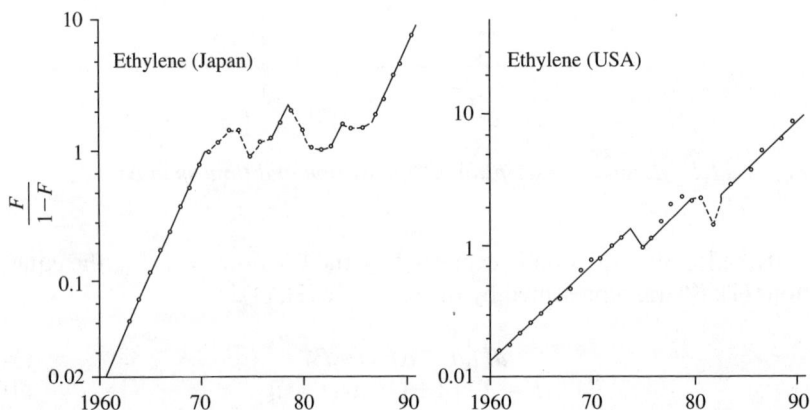

Figure 11.2   Diffusion trajectories of ethylene (petrochemicals)

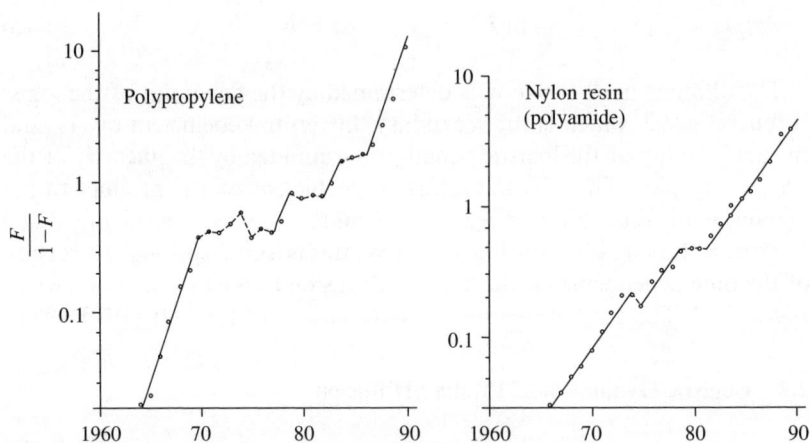

Figure 11.3   Diffusion trajectories of plastics

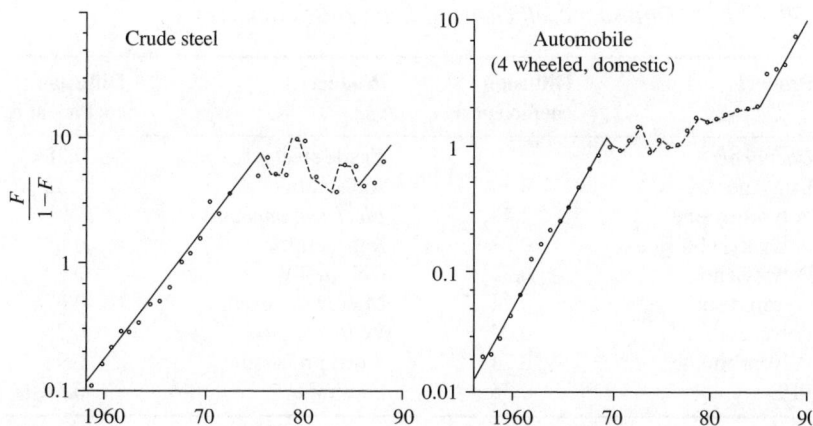

Figure 11.4 Diffusion trajectories of crude steel and automobile

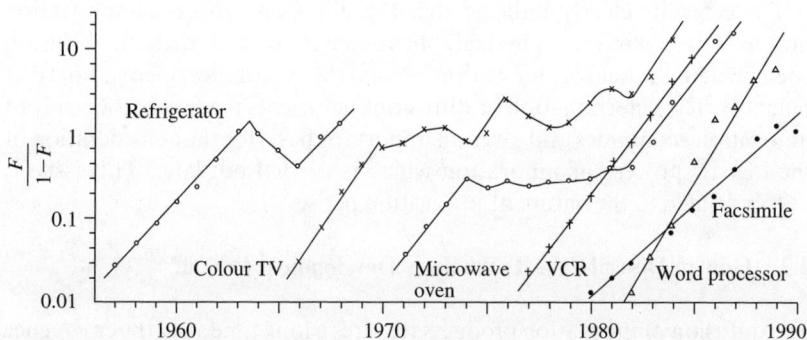

Figure 11.5 Diffusion trajectories of electrical appliances

These results clearly indicate that:

1. The diffusion of new products obeys a simple logistic equation during sound economic conditions;
2. The diffusion is easily disturbed by economic turbulence, such as recessions and wars, and sometimes the demand of products during turbulence is sufficiently reduced as to dissociate it from the locus of the logistic equation;
3. It is noteworthy that after the recession the diffusion of the product resumes and takes up the same slope of the logistic curve as before the recession. This strongly supports the fact that the diffusion of a product has its own inherent trajectory with a definite diffusion coefficient.

Table 11.1   Diffusion coefficients of innovation products*

Product	Diffusion coefficient α	Product	Diffusion coefficient α
Chemicals		Crude steel	0.28
Ethylene	0.39	Automobile	0.32
Polypropylene	0.49	*Electrical appliances*	
Polyvinyl chloride	0.23	Refrigerator	0.65
Polystylene	0.37	Colour TV	0.82
Nyron resin	0.24	Microwave oven	0.67
Polyacetal	0.27	VCR	0.73
Polycarbonate	0.24	Word processor	0.94
PPE	0.35	Facsimile	0.51

* Japanese market.

These results clearly indicate that the diffusion process of innovation products is a kind of physical phenomenon with a definite diffusion coefficient in the field of innovation beyond the occurrence of economic turbulences. The determination of diffusion coefficients is a first in the study of innovation economics and gives an important basis for the consideration of the logistic process of innovation which is carried out later. This issue is deeply related to the nature of innovation per se.

2.3   How to Describe the Technology Development Period?

The diffusion of innovation products requires a long time after the emergence of radical technology. That is, innovation products appear in the market and begin to diffuse after a long period of the development of the technology.

It is rather difficult to express concretely the states of technological development. By analysing the course of technological development, however, we can draw some implications: in the early stages of technology, it is hard to make rapid progress; after reaching a certain level of technology, it develops sharply; and at the final stage, the development of technology again slows down. This progress suggests a kind of sigmoid curve.

2.4   Bibliometric Analysis of Technology Development Period

The output of technological development is patents and new products. It is possible to describe the transition of the annual number of patent applications and annual developments of new products in the course of technology development. There are various case studies of the transition of patents and product development. Kaku (1986) intensively analysed the number of new products commercialized in the dyestuff industry in Germany and found a sharp distribution of the transition across the period 1850 to 1900. Achilladelis et al. (1990) described the transition of number of patents in plastics and pesticide industries after World War II and indicated sharp distribution of the trends in both cases. This chapter analyses these data by applying the Fisher–Pry plot and illustrates the results in Figure 11.6.

Figure 11.6 indicates sharp transitions and their Fisher–Pry plots clearly give straight lines which are direct evidence that they obey a logistic equation. Thus, a technology development period is illustrated as a logistic S-curve having a definite time span. Recently, Andersen (2001) analysed the transition for the number of US patents by classification for 100 years, from 1890 to 1990, and found that eight fields of chemicals, nine fields of electricals and electronics, 21 fields of mechanical engineering, two fields of transportation, and four fields of nonindustrials exhibited trends expressed by a logistic equation on the basis of regression analysis. She concluded that the innovation process was expressed by a logistic equation as judged by a bibliometric analysis of patents.

From the point of view of knowledge development, Marchetti (1979, 1980) and Modis (1992) indicate that various human activities, for example

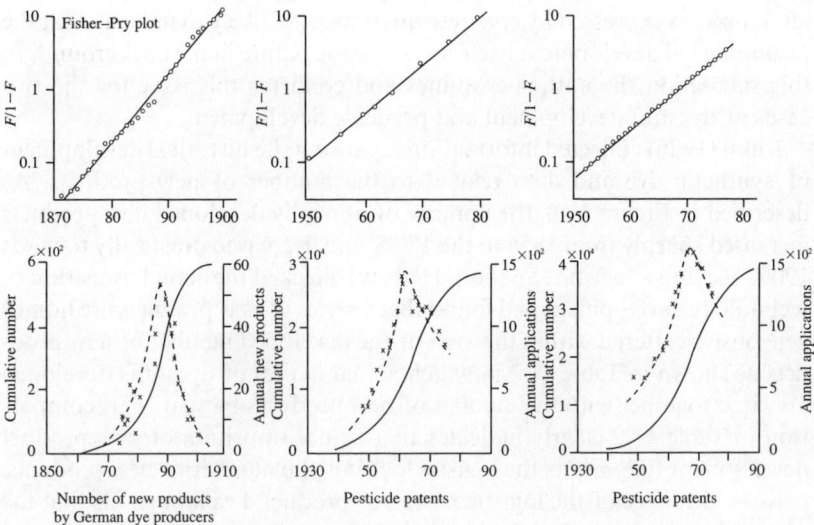

Source: Data from Achilladelis et al. (1990) and Kaku (1986).

Figure 11.6 Logistic nature of technology development

the music masterpieces of composers and the scientific papers of great scholars, can be expressed by a sigmoid curve. Marchetti (1980) also disclosed that the transition of various technological developments was expressed by a logistic equation; that is, the efficiencies of steam engines, generation of electric power, electric lamps, and the ammonia production process. He also pointed out that the number of chemical elements discovered was described by a logistic equation in the course of the discovery. These findings strongly support the logistic nature of technological development.

2.5   Evidence of the Logistic Nature of Technology Development

All of the above innovation processes are described by concrete quantities such as numbers of patents, numbers of new products, and some physical quantities. A problem is whether the real entity of technological development itself, neither the number of patents nor physical properties of technology, can be expressed by a logistic equation or not. Technological development is a phenomenon of knowledge transfer from person to person in the field of human society. This is a kind of discrete behaviour. The knowledge of technology, however, spreads over the relevant technological community and develops as if the knowledge continuously covers the entire field. Results of technological developments described above are only fruits materialized by such activity. The sigmoid curves of the innovation process expressed by concrete quantities are likely to indicate that the technological development itself has a logistic nature in the background. In this subsection the author examines and confirms this issue for the two cases of dyestuff development and pesticide development.

Kaku (1986) collected information regarding the historical development of synthetic dye and data related to the number of new products. As described in Figure 11.6, the number of annually developed new products increased sharply from 1856 to the 1880s and decreased drastically towards 1900, obeying a logistic equation. Here, we checked the actual transition of technology development and found that a series of new products are homogeneously scattered within the area of the developed number of new products as shown in Table 11.2 in which actual names of dyestuffs developed are cited together with the number of new products by year. This comparison in Table 11.2 clearly indicates that actual transition of new product development fully meets the logistic locus of the number of new products.

More evidence of the logistic nature of product development is that the actual distribution of individual pesticide products is scattered around the logistic S-curve of the number of patents of pesticides as shown in Figure 11.7.

These data reflect the fact that the technological development itself is dis-

Table 11.2   Product development and number of new products in the dyestuff industry in Germany

Year	Developed dyestuff	Number of new products
1856	Mauve	1
1859	Fuchsine	1
1863	Bismarck brown	1
1868	Alizalin	1
1871		5
1973	Eosine	2
1875		12
1877	Methylene blue	10
1878		20
1879		18
1880	Indigo	5
1883		23
1884	Congo red	10
1886		33
1895		22
1897	Indigo by naphthalene process	8
1899		8
1900		3
1901	Indanthrene	2

Source: Data from Kaku (1986).

tributed in the same locus as those described by concrete quantities of innovation and can be concluded to have a logistic nature. That is, the bunch of the discrete facts of developed technologies directly corresponds to the real locus of technological development. Thus, the logistic curve of a technological development can be determined by the concrete bunch of developed technologies and the size and positioning of the curve is determined by the time span of the bunch. Of course, the vertical axis corresponds strictly to the level of the technology or a degree of maturity. The nonlinear nature of innovation indicates that there is no technology progress before or after the development time span because of the breakthrough of the origin and the maturation of technology after the time span. It is a very interesting and important implication that all inventions of key technologies are gathered within a definite time span to make a bunch. A technology trajectory can be easily identified by checking how inventions are gathered during a definite time span and the time span of the trajectory can be determined by knowing the first and last core inventions in a bunch without measuring any quantitative mass.

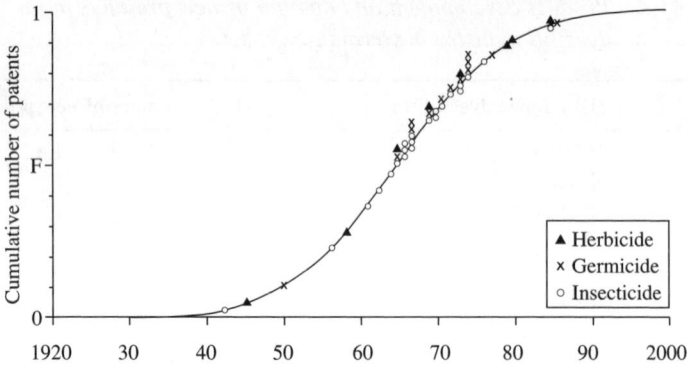

Source: Data from Sumitomo Chemical Co. (2002).

Note: The actual names of new pesticide products per annum were provided by courtesy of the Pesticide Division, Sumitomo Chemical Company Ltd., Japan and marked on the S-curve of annual developed numbers of pesticides.

Figure 11.7   Development of pesticides on patent distribution curve

2.6   How to Determine the Trajectory

It is important to know how to determine the trajectory. The actual method of determining the trajectory should be defined. As described above, if the measurement of the trajectory is carried out in the form of a Fisher–Pry plot, the slope of the straight line corresponds to the time span in the expression of the normalized scale of the vertical axis. As the logistic curve spreads from minus infinity to plus infinity, we have to decide what the time span is. We define the time span as the interval from $F = 0.1$ to $F = 0.9$. Technologies and new products emerge discretely and not continuously. This discrete phenomenon makes it possible to determine the time span. Thus, if we look directly at the bunch of technologies arranged in order of invention along the trajectory, we can define the first technology as $F = 0.1$ and the last one as $F = 0.9$. This kind of expression of time span has been adopted already by Marchetti (1979, 1980, 1988; Marchetti et al., 1996). It seems to be rather approximate but the actual determination was not so complicated to do and was successfully achieved for more than 40 innovations as shown in the next section. These results certainly reflect the non-linear nature of the trajectory.

Thus, an actual determination method is illustrated in the case of the innovation paradigm for electronics as shown in Table 11.3 and Figure 11.8.

According to the chronicle of electronics technologies from the 18th

Table 11.3   Elements of technology trajectory of electronics

Year	Inventor	Invention	Trajectory
1948	W. H. Brattain, J. Bardeen	Function of point contact transistor	F=0.1
1948	W. B. Shockley	Patent of p-n junction transistor	
1949	W. B. Shockley, G. L. Pearson, M. Sparks	Establishment of p-n junction concept	
1949	W. B. Shockley	Theory of p-n junction transistor	
1951	W. B. Shockley, M. Sparks, G. K. Teal	Completion of p-n junction transistor	
1952	W. B. Shockley	Concept of unipolar transistor	
1953	G. C. Dacey, I. M. Ross	Unipolar field effect transistor	Trajectory 1948–1973
1954	H. Krömer	Drift type transistor	$\Delta \tau = 25$
1956	Charles A. Lee	Diffused base transistor	
1959	Jack S. Kilby	Invention of solid state circuit	
1959	Jean A. Hoerni	Silicon planar transistor	
1960	Dawori Kaling, M. M. Atalla	MOS transistor	
1961	R. W. Noyce	Silicon planar monolithic integrated circuit	
1963	F. M. Wanlass, C. T. Sah	complementary metal oxide semi-conductor transistor	
1965	Texas Instruments	Schottky-cramped transistor	
1967	John T. Wallmark	metal-nitride-oxide-semi-conductor, Si-nitride-oxide semiconductor	
1971	Daglass L. Benzer	ISO planar technology	
1972	T. H. Philip Chang	Electron beam stepper	
1973	IBM	Submicron lithography	F=0.9

Source: Data from Kisaka (2001) and some others.

century to the present (Kisaka, 2001), it can be easily summarized that the origin of the technology trajectory was the discovery of the transistor by Shockley et al. in 1948 and the last core technology was the completion of

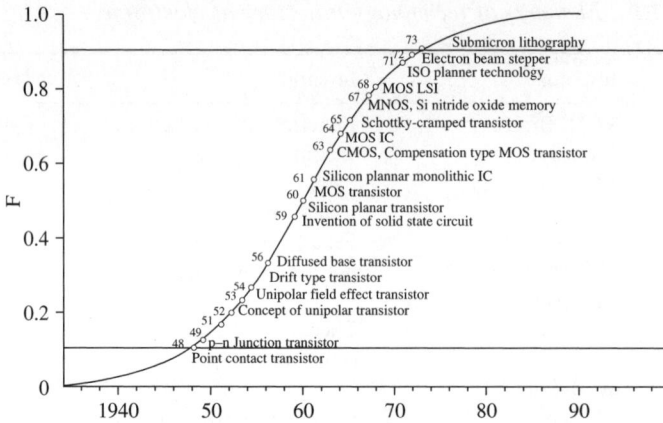

Figure 11.8   Identification of technology trajectory of electronics

submicron-lithography by IBM in 1973. Table 11.3 shows a series of core technologies that were developed almost every year. The time span is simultaneously recognized as 25 years and the S-curve is instantly described as shown in Figure 11.8. Each technology is placed on the curve by name and year.

2.7   Determination of Time Span for Innovation Technologies

In order to confirm the finiteness of technological development, the author has determined the time span of technological developments for more than 40 innovations since the Industrial Revolution to the present day as shown in Figure 11.9. These bunches of technologies are identified by various data from McNeil (1990), Yuasa (1989), other chronological technology handbooks and textbooks. The identification of trajectory elements often requires expert knowledge and the results were often confirmed by experts from the relevant disciplines. The author has collected these data for more than ten years. It was, however, surprising to find that after the elements were confirmed, determining the time span was not so difficult, as the development of innovative technologies clearly lined up within a definite time span. This seems to be the first time an abstract phenomenon as technology development is described by a concrete equation. This success is certainly ascribed to the nonlinear nature of innovation and the determination only requires the time span of the spread of technologies to be measured without cognition of the vertical axis, which is normalized as a fraction of scale towards saturation at infinity.

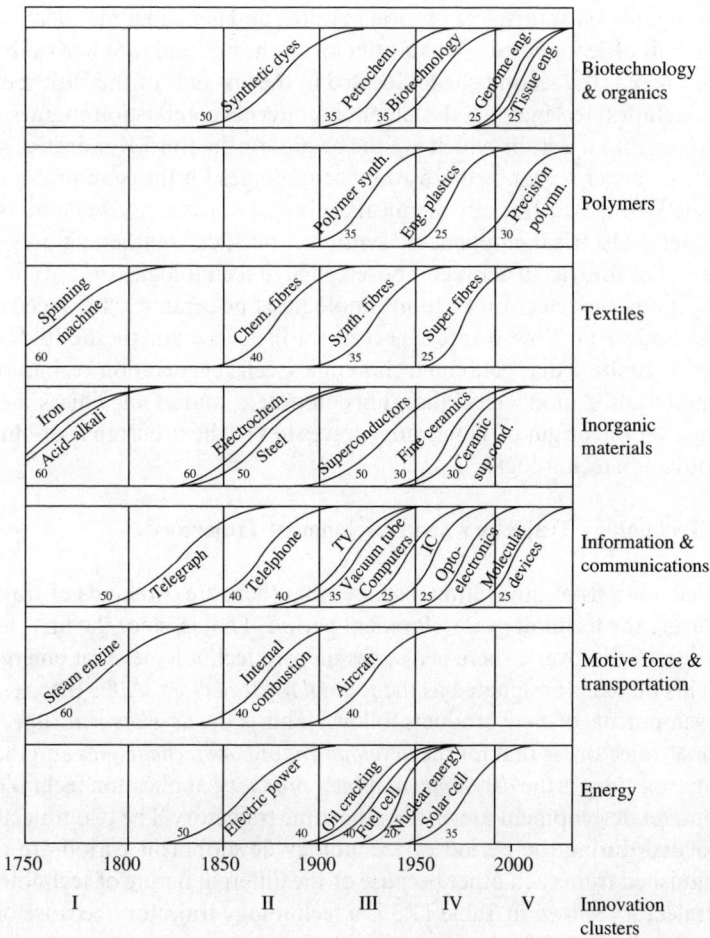

Figure 11.9   Determination of time span of technology development period

Figure 11.9 shows that the time span of developed technologies was around 60 years at the age of the Industrial Revolution and is now shortened to 25 to 30 years. Kondratiev (1926) proposed a concept of a long wave of business cycles and Schumpeter (1939) ascribed the business cycles to the cluster of innovations. Hirooka (2002) provided direct evidence for Shumpeter's postulation and found that most innovations which had a major impact on the economic infrastructures were intensively gathered on the upswing of Kondratiev waves. Related to this finding, Figure 11.9 clearly shows clusters of innovation trajectories which seem to be the origin of business cycles. That is, corresponding to the Industrial Revolution, there is a cluster of spinning machines, steam engines, blast furnaces for iron making, and acid/alkaline chemicals in the last half of 18th century. This cluster forms the first and second Kondratiev waves. There is the second cluster located in the last half of the 19th century, which includes steel making, the telephone, internal combustion engines, electric power, and dyestuffs which are the origin for the third Kondratiev wave. The third cluster is that of innovation technologies for the economic growth after the World War II: petrochemicals, oil cracking, aircraft, vacuum tubes, computers, electrical appliances, synthetic plastics, synthetic fibres, and rubbers. The fourth cluster is composed of high technologies for the contemporary industries: electronics, biotechnologies, fine ceramics, advanced composites, and so on. Now, we are expecting the fifth wave and these could be the cluster of multimedia, genome technologies, cell regeneration technologies, nanomaterials, nanodevices, quantum computers, and so on. This is the first finding that the origin of Kondratiev waves lies in the occurrence of clusters of innovation technologies.

2.8   Technology Trajectory and Development Trajectory

Detailed analysis of innovation discloses that there are two kinds of trajectories during the technology development period. That is, since the first radical invention or discovery, there is a series of core technologies that emerges in line. This bunch is designated as the *technology trajectory*. After that, a series of developments of new products follows. This is the *development trajectory*. The first trajectory is that for the *development of core technologies* and the following trajectory is the *development of new products*; application technologies and market development are also on the same trajectory. The two trajectories formulated during the period of technology development period are easily distinguished from each other because of the different nature of technologies. The trajectory shown in Table 11.3 is a technology trajectory because of the core technologies, while the development of new dyestuff products as given in Table 11.2 and pesticides as shown in Figure 11.7 are development trajectories because they describe the development of new products. Most of the trajectories depicted in Figure 11.9 are technology trajectories.

3.   IDENTIFICATION OF AN INNOVATION PARADIGM

3.1   Correlation of Technology Trajectory and Diffusion Trajectory

An innovation paradigm is composed of three trajectories: technology, development, and diffusion trajectories, in this order. Interestingly, the diffusion trajectory is seen to begin just after the technology trajectory is almost complete, like a kind of cascade junction. This relation is an empirical result, but it is important to describe the innovation paradigm and to gain foresight in technology. Figures 11.10 and 11.11 depict typical examples of the interrelation between the actual technology trajectory and the diffusion trajectory in various innovations such as energy development, motive powers, and information and communication technologies. All of these paradigms clearly indicate that the technology trajectory joins with the diffusion trajectory in a cascade fashion. The source of data is the same as in Figure 11.9 and industrial statistics.

3.2 Structure of an Innovation Paradigm

Now, let us describe the structure of the innovation paradigm consisting of technology, development, and diffusion trajectories. The following paragraphs describe the electronics, biotechnology, and synthetic dyestuff paradigm.

The technology trajectory of the electronics innovation paradigm has been introduced in Table 11.3 and Figure 11.8. The original invention is the discovery of the transistor by Shockley et al. in 1948 and then core inventions of integrated circuits by Kilby and Noyce, MOS IC, and steppers followed. In 1973 IBM finally completed the technology of submicron-lithography. This indicates that the time span of the technology trajectory of

(1) Energy and motive powers

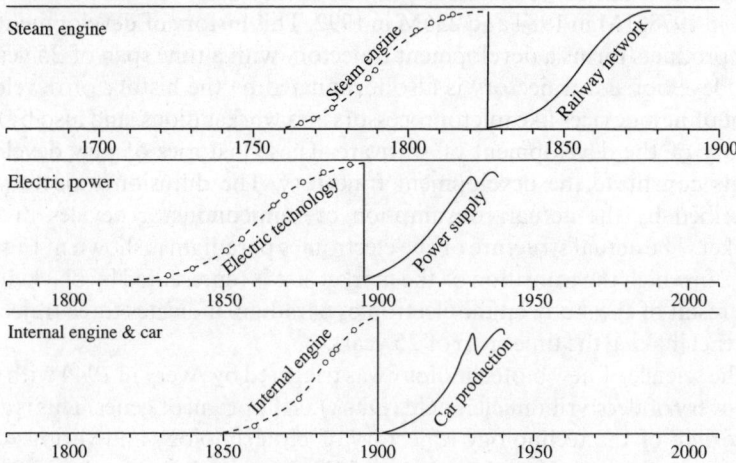

Figure 11.10   Cascade of technology and diffusion trajectories

(2) Information and communications technologies

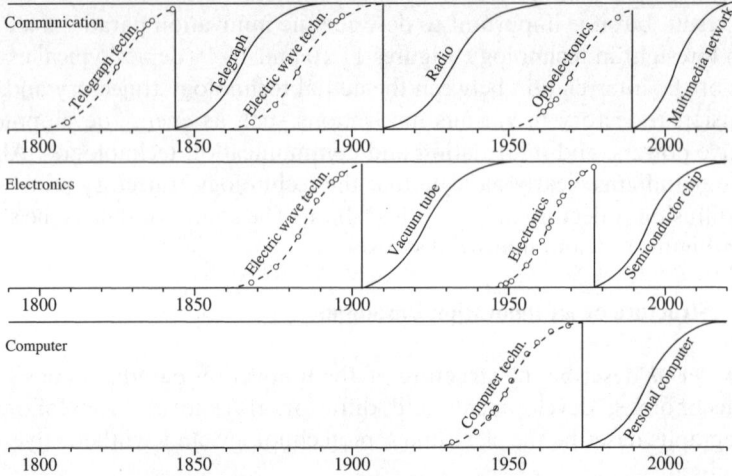

Figure 11.11   Cascade of technology and diffusion trajectories

25 years from 1948 to 1973 and we can illustrate the technology trajectory in such a way. The development trajectory is the history of development of integrated circuits which started from the completion of submicron-lithography in 1973. The growth of DRAM formulates the development trajectory in which the degree of integration is advanced four times every three to four years, known as Moore's law. The progress is shown as 16K DRAM in 1975, 64K in 1978, 1M in 1984 and 256M in 1992. This history of development of new products forms a development trajectory with a time span of 25 years. The development trajectory is also formulated by the history of development of new devices like microprocessors and workstations, and also by the history of the development of software. These histories of new developments constitute the development trajectory. The diffusion trajectory is described by the actual consumption of semiconductor devices in the market. The actual structure of the electronics paradigm is shown in Figure 11.12 in which the transition of the market size is represented by black dots. The result of the study on the electronics paradigm indicates three trajectories each having the time span of 25 years.

The science of new biotechnology was triggered by Avery in 1944 with the discovery of deoxyribonucleic acid (DNA) as the origin of genes. This is also the origin of the technology trajectory of biotechnology and various discoveries such as the helical structure of DNA, central dogma, decoding of genes, and recombinant DNA, and the synthesis of insulin by E. coli (colon

Submicron lithography 73
Electron beam stepper 72
ISO planar technology patent 71
MOS LSI 68
MNOS (Si-nitride oxide) memory 67
Schottky-cramped transistor 65
MOS • IC 64
CMOS (compensation type MOS) 63
Silicon planar integrated circuit 61
MOS transistor patent 60
Silicon planar transistor 59
Invention of solid state circuit 59
Diffused base transistor 56
Drift type transistor 54
Junction FET 53
Idea of unipolar transistor 52
P–N junction transistor 51
P–N junction theory 50
Point contact transistor 48 49
P–N junction

Microprocessor SPARC (SUN)
Macintosh (Apple) 82 1M 84
Workstation (SUN-1) 82 82
256K
64K DRAM 78 79 BDS (UNIX)
RISC microprocessor 75 ALTO (XEROX)
Personal computor 74 75 16K
Intel 8008 73 4K Basic (Microsoft)
69 UNIX (AT&T)

1G 4G 95
Windows95
256M
64M
90 Windows (Microsoft)
89
88 16M
87 UNIX International
4M 86
Lotus-1, 2/3
MS-DOS (Microsoft)

Development trajectory
Technology trajectory
Diffusion trajectory

F

$\Delta\tau = 25$ $\Delta\tau = 25$ $\Delta\tau = 25$

1930   40   50   60   70   80   90   2000   10

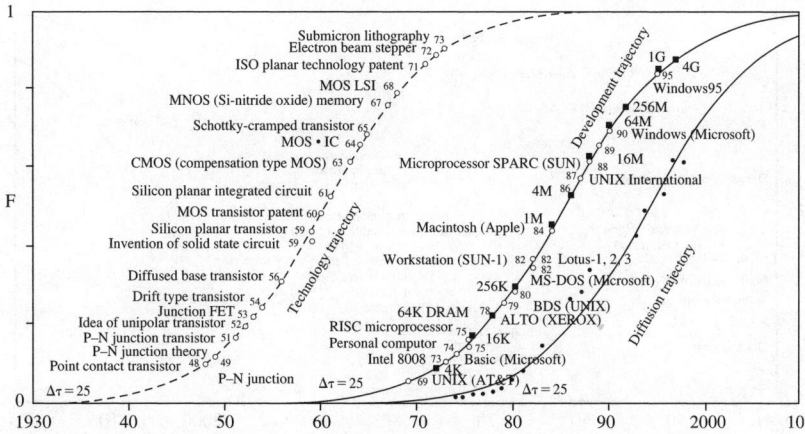

Source: Data from Kisaka (2001), MITI (2000), Shimura (1992), and others.

Figure 11.12   Innovation paradigm of electronics

bacillus), which together constitute the technology trajectory that lasted for 35 years. The development trajectory started with the development for the commercialization of human insulin by E. coli and the various biotechnology products that developed along the trajectory. The biotechnology paradigm is shown in Figure 11.13.

The first synthetic dyestuff was invented by Perkin in 1856 at the London College of Chemistry and until the early 20th century many more synthetic dyestuffs were developed on as shown in Figure 11.6 and Table 11.2. Such progress is described as the development trajectory. The technology trajectory corresponds to the development trajectory and is the history of organic chemistry, mostly at universities, starting from the identification of organic substances by Wöhler to the establishment of organic chemistry by Kekulé. The actual sales of dyestuffs describe the diffusion trajectory. The innovation paradigm of synthetic dyestuffs is given by Figure 11.14. The time span of each of these trajectories is 40 years.

4.   IMPLICATIONS OF THE LOGISTIC DYNAMISM OF INNOVATION

The innovation paradigm consists of three trajectories and has a logistic nature which indicates that every trajectory reaches an ultimate level of saturation within a definite time period. Each trajectory plays its own role

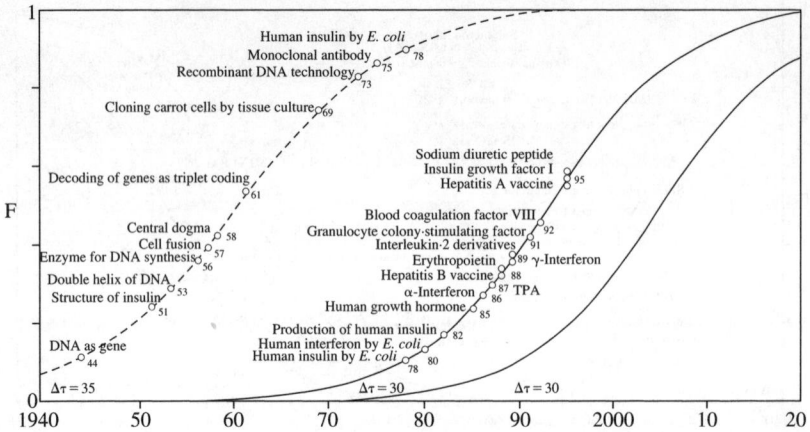

Source: Data from Lewin (1983), JBA (2003), Walker & Gingold (1985) and others.

Figure 11.13   Innovation paradigm of biotechnology

and the actors involved are also different. In this sense, knowledge transfer across the trajectories and within a trajectory as well is also very important. The technology trajectory establishes core technologies, the development trajectory develops new products and the diffusion trajectory focuses on sales. Each trajectory has its own role. Among these trajectories, the development trajectory is the most important in the sense of innovation. In this section, some characteristics and roles of the development trajectory are discussed.

4.1   Role of Universities in the Innovation Paradigm

The technology trajectory is often composed of the fruits of basic research that is mostly carried out at universities, which means scientific research. The technology trajectory of synthetic dyestuffs is the outcome of pure research in organic chemistry that was carried out at universities. Biotechnology was based through basic research into gene science at universities, which started with the discovery of genes by Avery and ended with the great achievement of recombinant DNA by Cohen and Boyer. The technology trajectory of polymer science is also a kind of scientific trajectory in which research was mostly carried out at universities, starting with the concept of macromolecules by Staudinger. The technology trajectory of electronics is formulated mostly by research at firms but rich in basic flavour. Further, many universities were involved, in particular Stanford University in establishing elec-

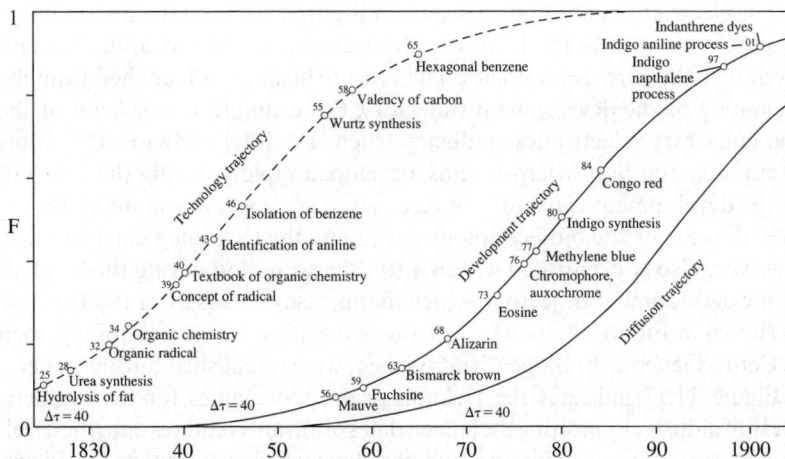

Source: Data from Aftalion (1991), Kaku (1986) and others.

Figure 11.14   Innovation paradigm of synthetic dyestuffs

tronics technologies. The development trajectory is often the field of technology transfer from universities to firms. In the case of synthetic dyestuffs,
the first invention of a dyestuff was carried out at the Royal College of
Chemistry in London. Most dyestuff products, however, were invented by a
collaboration of universities and firms in Germany. Professor Adolf Bayer,
of the Berlin Institute of Technology played a key role and research staff at
BASF and Hoechst, chemical companies in Germany, collaborated with
him. The development trajectory of biotechnology is formulated by the
development of new products resulting from the technology transfer from
universities to firms or by spinout of people from universities.

4.2   Technological Opportunities and Timing of Venture Business

In the course of the innovation process, the most important occasion is the
emergence of venture business. The logistic dynamism of innovation clearly
illustrates the importance of technological opportunities and the timing of
venture businesses. Within the technology trajectory, core technologies are
developed and the technological opportunities are accumulated. At the
beginning of the development trajectory, venture business is launched. This
kind of relationship is examined for several cases by plotting the set-up time
of venture businesses on the development trajectory.

The electronics paradigm is shown in Figure 11.12 and the development trajectory is given as the locus of advancement of DRAM after the completion of the core technologies. The venture business is launched from the beginning of the development trajectory. For example, major firms of the contemporary electronics industry such as Intel, Microsoft, Apple Computer, and Sun Microsystems, developed rapidly during the first half of the development trajectory of electronics as shown in Figure 11.15.

In the case of the biotechnology paradigm, the technological opportunities were also concentrated within a 10–15 year period during the first half of the development trajectory, which in this case extends from 1980 to 2005 as shown in Figure 11.16. Most of the active firms in biotechnology, such as Cetus, Genentech, Biogen, and Amgen, were established during this era.

Figure 11.17 indicates the technological opportunities for the synthetic dyestuff industry by plotting the launch dates of various ventures onto the development trajectory of the innovation paradigm which is formulated in Figure 11.14. The first venture firm was established by Perkin just one year after he discovered mauve dye at the Royal College of Chemistry. The leading chemical companies of today such as BASF, Hoechst and Bayer were established around this time as small venture firms. The launching of venture businesses is concentrated in 10–15 years in the first half of the development trajectory.

These results clearly indicate that opportunities for venture businesses with the innovation paradigm are concentrated within a limited period

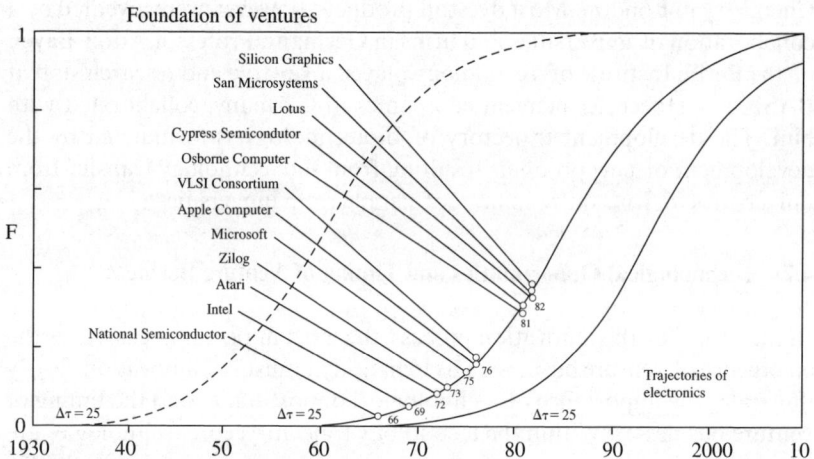

Source: Data from Nihon Keizai Shinbunsha (1996) and others.

Figure 11.15   Timing of venture business in electronics paradigm

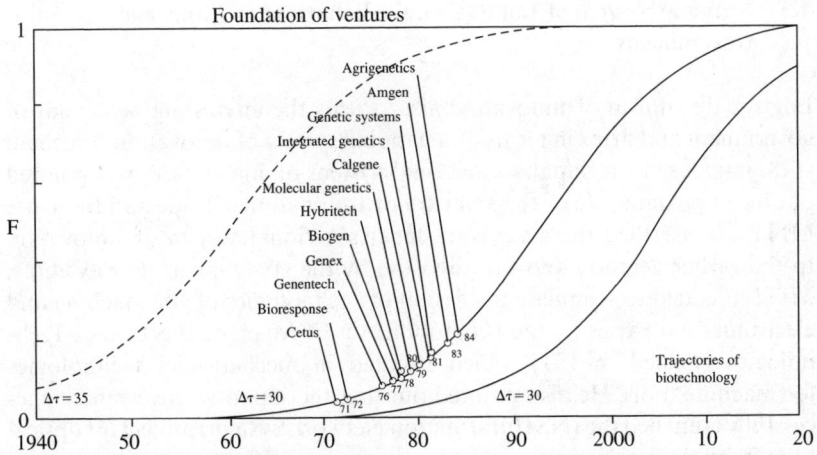

Source: Data from JBA (2003) and others.

Figure 11.16 Timing of venture business in biotechnology paradigm

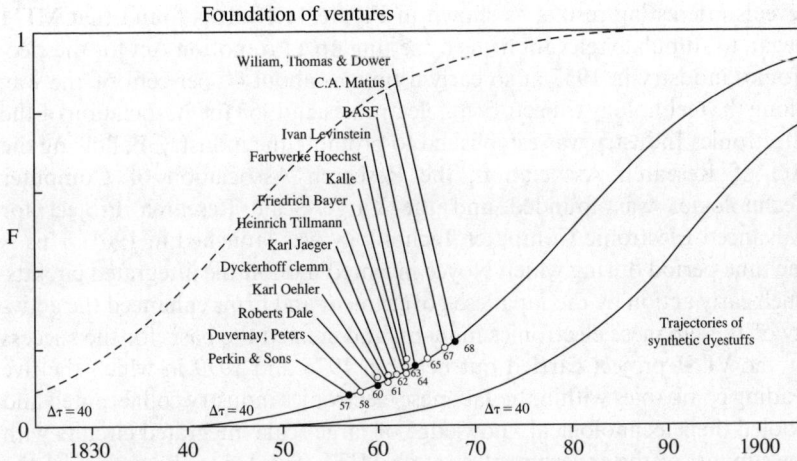

Source: Data from Kaku (1986).

Figure 11.17 Timing of venture business in dyestuff paradigm

during the first half of the development trajectory. Such common behaviour of venture business accords with what Schumpeter described as a bandwagon effect. Utterback (1994) also recognized a sharp increase of firms at the beginning of innovation diffusion.

4.3 National System of Innovation, the Behaviour of Firms and Governments

Logistic dynamism of innovation also reveals the interesting behaviour of government and firms in terms of a national system of innovation. Freeman (1987) analysed the Japanese national system of innovation and pointed out the important role of the Ministry of International Trade and Industry, MITI. He ascribed the success of Japan's national system of innovation to the earlier decision and future vision by the government. For example, MITI intended to stimulate the fusion of technologies of the machine and electronics industries by the Promotion Act of Machine/Electronics Technologies effected in 1971 which resulted in mechatronics technologies for machine tools. He also pointed out that the Japanese government successfully launched the INS (Information Network System) project for optical fibre technologies at a very early stage on the basis of clear perspectives.

We analysed the role of MITI and other relevant firms in the electronics businesses in relation to the innovation dynamism. The innovation dynamism of electronics is described by three trajectories as shown in Figure 11.12. The timing of actions effected by MITI is put on these trajectories and reveals interesting results as shown in Figure 11.18. It is found that MITI began to stimulate relevant firms by setting up a Promotion Act for the electronics industry in 1957 at so early a time as about 40 per cent of the way along the technology trajectory for electronics; in 1958 the Association of the Electronics Industry was established to promote the industry. Following the Act of Research Association, the Research Association of Computer Technologies was founded and the Large Scale Research Project for Advanced Electronic Computer Technology was launched in 1961. This is the time period during which Noyce invented monolithic integrated circuits. Such early action by the Japanese government and firms enhanced the activity of the Japanese electronics industry and created the basis for the success of the VLSI project carried out between 1976 and 1979 in which the five leading companies within the Japanese electronics industry collaborated and pooled their technological knowledge of large scale integrated circuits with the support of the government through MITI. This led to the success of the 64k DRAM in Japan in 1981, which was earlier than in the USA. The timing of these actions can be clearly understood by putting them on the electronics trajectories in Figure 11.18 where the VLSI project is placed on the development trajectory because the actions are those for product development. It is noteworthy that the timing of the VLSI project is included only in the area of highly concentrated technological opportunities for venture businesses as pointed out in section 4.2. The logistic analysis is an effective way to understand this kind of situation.

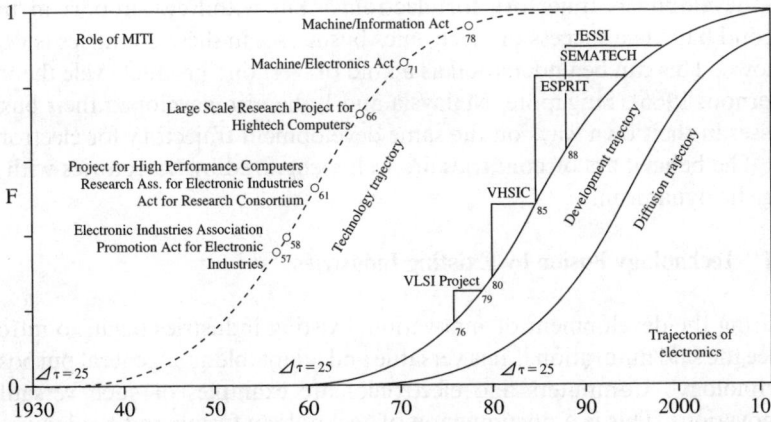

Source: Data from Freeman (1987), HTSRG (1990) and others.

Figure 11.18 Role of government and collaboration projects for electronics

In the USA, collaboration of research between companies belonging to the same industry had been prohibited by the Anti-Trust Law. After the Presidential Message on Innovation by President Carter, the USA had softened their regulations and allowed firms to collaborate with each other. In response to the Japanese collaborative movement, represented by the VLSI project, the US Department of Defense in 1980 launched the five year long VHSIC project which encouraged leading electronics companies to develop very high speed and sophisticated integrated circuits; the Collaboration Research Organization SEMATECH was established in 1988. The European Community followed these movements and set up the European collaboration project ESPRIT in 1985 as a ten year long project to develop high level integrated circuits in order to catch up with the Japanese and US activities. Figure 11.18 depicts these actions on the development trajectory for electronics and we can compare the timing of these projects at a glance. The national system of innovation concept has now become popular throughout the world.

The Japanese electronics industry successfully developed their business with the aid of a national system of innovation. Following the so-called 'flying geese model', Asian countries followed on Japanese success. It is interesting to see that the businesses within every country have moved on the same development trajectory for electronics. This means that electronics businesses spread worldwide at the same time. Korean and Taiwanese governments actively supported electronics businesses in the first half of the development trajectory for electronics but withdrew support in the second half. The success of electronics businesses in these countries is well known. This can be understood as a kind of Vernon's product cycle theory (Vernon, 1966). Singapore, Malaysia and India also developed their businesses in their own ways on the same development trajectory for electronics. The behaviours of countries are well recognized by trajectories with a logistic dynamism.

4.4   Technology Fusion by Existing Industries

During the development of innovation, existing industries begin to introduce the said innovation if it is versatile and adaptable as a 'general purpose technology'. Computers and electronics are examples of such versatile innovations. This is a phenomenon of technology fusion and the locus of technology introduction makes such a fusion a trajectory. The introduction of computer technologies by the steel, cement, chemical, and railway industries and in flight simulators was studied (Hirooka, 1998b) and it was found that the fusion trajectories for the introduction of computers was located between the development and the diffusion trajectories for computers. In the case of introducing computers only for users, computers were adopted along the diffusion trajectory but, when computer technologies were introduced to aid in the development of their products, the fusion trajectory by adoption was closely associated with the development trajectory.

In the case of electronics technologies, it is interesting to note that the machine tool, camera, automobiles, and airplane engine industries adopted the innovation technologies as development processes and the fusion trajectories developed along the development trajectory for electronics itself. Figure 11.19 illustrates the technology fusion of electronics in the car industry.

5.   NEW INSIGHT INTO THE ECONOMICS OF TECHNOLOGICAL CHANGE

This chapter is the first attempt to describe the whole figure of the innovation paradigm using three trajectories. In particular, it is interesting to express an abstract matter such as technological development by a visual concrete curve. One of the important characteristics of this analysis is to express the progress of innovation by the use of trajectories having a nonlinear nature. While the diffusion of innovation products has been well known to obey a logistic equation, it is interesting to find that the transition of technological development can be described by a logistic equation as well. This kind of

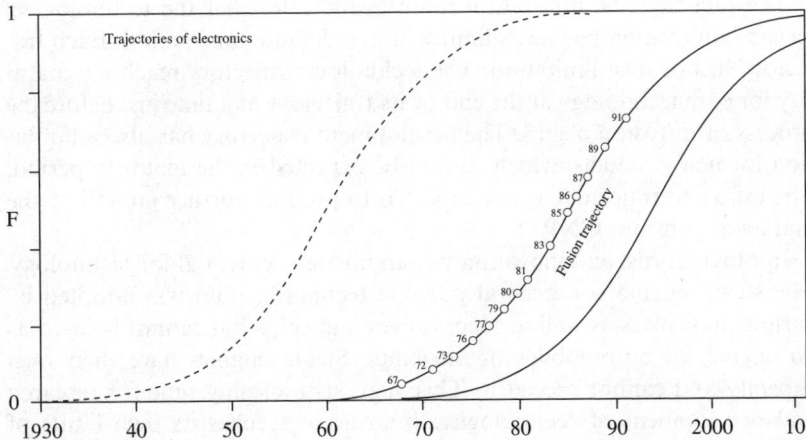

Source: Data from Takano (1997).

Figure 11.19 Technology fusion of electronics by car industry

derivation has revealed various interesting features and the concept of innovation is further interpreted from different angles.

According to the present treatment of the innovation paradigm using trajectories, there are three features to be pointed out. One is the nonlinear nature of innovation, which means that each of these trajectories has an ultimate saturation level. This implies that timing during innovation is very crucial. The second feature is the exogenous nature of the technology trajectory, introducing a trigger for innovation, which means that the origin of innovation often comes from outside the economy, in particular, from universities. The third feature is the importance of knowledge transfer among intra- and inter-trajectories. The concept of a field for knowledge transfer should be noted because knowledge transfer is essential for the progress of innovation.

5.1 Nonlinearity of Innovation

The nonlinear nature of innovation provides the phenomenon for the trajectory to reach a level of saturation within a definite time period. This makes it possible to identify the trajectory only by determining the time span of the trajectory. It also indicates the timing of various phenomena for innovation, as shown by the timing of venture business, the timing of government policy, and the timing of technology fusion. Thus, innovation should be recognized as a function of time.

The fact that the innovation matures indicates that the technological level of innovation reaches a limit within a definite time period. Each trajectory has its own limitation. The technology trajectory reaches a maturity for core technology at the end of its trajectory at a time just before the process of diffusion begins. The development trajectory has also a limitation for new products, which cannot be expected in the maturity period. The diffusion trajectory is not expected to produce further growth of the market at constant GNP.

In other words, an innovation has an inherent potential for technology. The steam engine is a general purpose technology and was adopted by various industries as well as locomotives and ships but cannot be used as an engine for automobiles or airplanes. Steam engines have their own capacity and cannot exceed it. This suggests a closing time for research and development of technologies. Innovation reaches its own limits of abilities within a definite time and retains it afterwards. This kind of phenomenon resembles the evolution of a living organism. For instance, the coelacanth evolved 350 million years ago and its shape has remained constant since.

In the course of the diffusion process for innovation, the added value of innovation products increases and reaches a limit at maturity. The impact of innovation on the economy depends on its inherent potential and capabilities. The economic impact index should be evaluated as a product of unit value added and population receiving the benefit.

The time span for innovation has shortened over time. How short a span can we expect in the future? Now that the speed of information transmission is considerably enhanced, the time span must depend on the capabilities of human brains to process this information. As this cannot be expected to be improved, the time span of innovation might not be shortened below some threshold, say 20 years.

5.2 Wider Concept of Innovation with Exogeneity

The concept of an innovation paradigm having three trajectories has not been recognized so far and most discussion of innovation has been carried out within a scheme of two trajectories of development and diffusion to describe the behaviour of firms. Of course, many discussions of innovation have dealt with the activities of universities for basic research but, there has not been any recognition of a technology trajectory as an important element. Basic research at universities is carried out independently of the economy and has a kind of exogeneity element to it.

According to our analysis, core technologies are frequently created in universities and end at the beginning of the development trajectory. This means that almost all core technologies have been completed before firms are involved by technology transfer from universities. Recently, there has been a tendency to directly commercialize the scientific knowledge created at universities and venture businesses are easily launched. From this point of view, earlier catch-up of created knowledge results in more successful business. Firms should be involved in the innovation process earlier on in the technology trajectory, and preferably collaborate with universities. Rosenberg (2001) discussed this matter and pointed out the importance of the endogeneity of university research.

The trajectories of the innovation paradigm are always arranged in the order of technology, development and diffusion. This means that the basic research, development of new products, and market formation occur in this order. This is the concept of the linear model of innovation which, again, is the accepted feedback model. This is ascribed to the introduction of the technology trajectory as an essential element. A feedback model has been discussed as a phenomenon around the development and diffusion trajectories. In this case, knowledge transfer between trajectories is essential and the feedback effect is understood to be an important factor. This is the feedback model versus the linear model. In our study, we would also like to emphasize the importance of knowledge transfer among intra- and inter-trajectories which is the same as the feedback model. We should, however, consider the technology trajectory before the industrial innovation field consisting of development and diffusion trajectories. Such a relation results in a revival of the linear model.

Nelson and Winter (1982) skilfully described the innovative behaviour of industrial firms as the same process as the evolution of life. In this evolutionary theory of economics, they discussed the behaviour of firms as routines which play the same roles as genes. Routines evolve through regeneration, mutation, and survival of the fittest. Firms having supreme routines survive in the struggle for existence and then natural selection proceeds.

We have two kinds of evolution in the organism: phylogenic evolution (macroevolution) and ontogenic species evolution (microevolution). The latter corresponds to the innovation which evolves along the innovation trajectories. This is the development of technologies which is a kind of 'technological evolution', such as the evolution of human beings from ape man, hominid, and to *Homo sapiens*. On the other side, the innovation paradigm itself jumps to another paradigm by evolution. This is a kind of phylogenic macroevolution which could be described as 'paradigm evolution' and corresponds to the bifurcation of mankind from apes. University research induces paradigm evolution and should be carefully watched in this sense.

5.3   Real Entity of Innovation and the Concept of a 'Field' for Knowledge Transfer

In the innovation paradigm, it is important that the knowledge in innovation is transferred within a trajectory and, especially, across trajectories. This kind of phenomenon is recognized as a 'field' of knowledge transfer formulated by the various actors involved in the trajectories in the human community.

The field of knowledge transfer in the innovation paradigm is actually a discrete system because of person-to-person transfer of knowledge. In a discrete system, the phenomenon should not be expressed by a logistic equation but could be described with a difference equation. It could, however, be permissible to apply a logistic equation to a system if a large number of people are involved, such as a social community, as a homogeneous system, according to a convention of continuity for mathematical description. The system of an innovation paradigm is usually applied to this kind of assumption. In the case of a discrete system, what corresponds to a logistic equation is, for example, May's mapping. (May, 1974). If we apply this difference equation to the actual diffusion of innovation products shown in Table 11.1, the order of diffusion coefficients is indicated to be in an ordinary condition in which the convention of continuity can be applied. We have, however, observed the existence of fractals in the innovation paradigm as shown in Figure 11.20, which illustrates the evolution of three paradigms of information technologies with development and diffusion trajectories, and the fractals of new products developed by innovation are represented by parallel bars across two trajectories. The development trajectory of electronics is also composed of a series of fractals of semiconductor devices in which each device develops within a three to four year period and steps up to the next. This clearly indicates the existence and composition of fractals along the development trajectory. Within these small fractal conditions, we might have to use difference equations like May's mapping because of a discrete phase.

Nelson and Winter (1982) describe how an innovation must be a kind of Markov chain system. That is, the evolutionary system of innovation progresses through a trial and error process and is a kind of stochastic one with high uncertainty. This is an acceptable consideration but the resulting paradigm has well defined trajectories as a deterministic process. There is no contradiction between the two phenomena because, in the course of innovation, there is much uncertainty according to a stochastic process but, the final states reach a deterministic logistic curve. This correlation is considered as a similar phenomenon to light, which has the two phases of stochastic probabilities and deterministic entity as a particle.

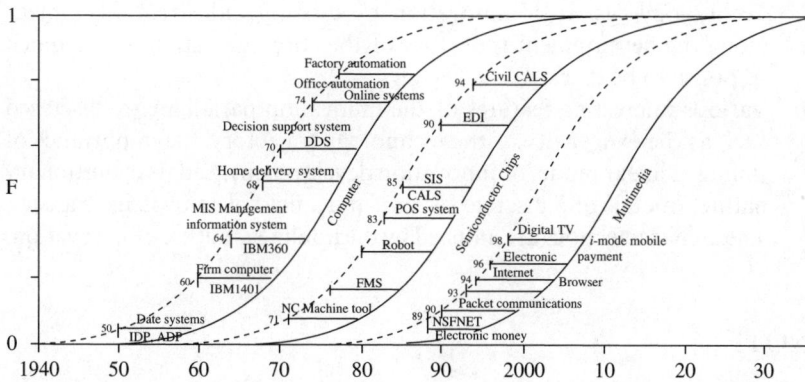

Figure 11.20   Evolution of information paradigms and fractals of elements

6.   CONCLUDING REMARKS

This chapter throws light on the technological development period and discloses various characteristic features of technological innovation. Important conclusions are as follows:

1.   The nonlinear nature of innovation is revealed and various relevant features are presented.
2.   Before the creation of a new market in the course of technological innovation, there is a long latent period of technological development.
3.   The technological development period consists of two phases; the development of core technologies and the development of new products. The former is designated as the 'technology trajectory' and the latter as the 'development trajectory'. Thus, an innovation paradigm consists of three trajectories of technology, development, and diffusion.
4.   Each trajectory has a nonlinear nature with a definite maturity level and can be expressed by a logistic equation.
5.   A trajectory can be determined by estimating the time span of a bunch of relevant technologies: the author adopts the interval from $F = 0.1$ to $F = 0.9$ in the vertical axis as the time span because almost all elements of a trajectory are included within this span.
6.   The technology trajectory represents a series of basic research activities carried out at universities and the development trajectory is formed by technology transfer from universities to firms.
7.   The technological opportunity for venture business: 'business chance', is limited to a rather narrow time period that is concentrated in the first half of the development trajectory.

8. Technology fusion of innovation by existing industries takes place along the development trajectory of the core innovation with a quick response to be digested.
9. Various interesting features of the innovation paradigm are disclosed such as the exogeneity of the technology trajectory, the importance of timing, a linear model of innovation development, and its evolutionary nature, fractals of a discrete system, and a field of knowledge transfer. These characteristics are induced by the nonlinear nature of innovation.

NOTE

1. Comments by Stan Metcalfe, Rob Coombs, Ian Miles are gratefully acknowledged.

REFERENCES

Achilladelis, B., A. Schwartzkopf and M. Cines (1990), 'The dynamics of techno-logical innovation: the case of chemical industry', *Research Policy*, **19** (1), 1–34.
Aftalion, F. (1991), *A History of the International Chemical Industry*, Philadelphia, PA: University of Pennsylvania Press.
Andersen, B. (2001), *Technological Change and the Evolution of Corporate Innovation*, Cheltenham, UK and Northampton, MA: Edward Elgar.
David, P. (1975), *Technical Choice, Innovation and Economic Growth*, Cambridge: Cambridge University Press.
Davies, S. (1979), *The Diffusion of Process Innovations*, Cambridge: Cambridge University Press.
Fisher, J. C. and R. H. Pry (1971), 'A simple substitution model of technological change', *Technological Forecasting and Social Change*, **3** (2), 75–88.
Freeman, C. (1987), *Technology Policy and Economic Performance: Lessons from Japan*, London: Pinter.
Griliches, Z. (1957), 'Hybrid corn: an explanation in the economics of technological change', *Econometrica*, **25** (4), 501–22.
Hirooka, M. (1998b), 'Dynamism of technological innovation and technology fusion towards complex system', *Journal of the University of Marketing and Distribution Sciences; Information, Economics and Management Science*, **7** (1), 49–70 (in Japanese).
Hirooka, M. (2002), 'Nonlinear dynamism of innovation and business cycles', Paper presented at the 9th Conference of the International J. A. Schumpeter Society, Florida.
Hirooka, M. and T. Hagiwara (1992), 'Characterization of diffusion trajectory of new products in the course of technological innovation', *Kobe University Economic Review*, **38**, 47–62.
High Technology Strategy Research Group (HTSRG) (1990), *Technological Competitiveness in Japan and the United States*, Tokyo: Nikkei Science Inc.
JBA (2003), data from Japan Bioindustry Association.

Kaku, S. (1986), *Introduction of the History of Chemical Industry in Germany*, Kyoto: Minerva Publishers.

Kisaka, S. (2001), *Chronicles of Science and Technology of Electronics*, Tokyo: Daily Industrial Newspapers Inc.

Kondratiev, N. D. (1926), 'Die Langen Wellen der Konjuktur', *Archiv fur Sozialwissenschaft und Sozialpolitik*, **56**, 573–606.

Lewin, B. (1983), *Genes*, New York: Wiley.

Mansfield, E. (1961), 'Technological change and the rate of imitation', *Econometrica*, **29** (4), 741–66.

Mansfield, E. (1963), 'Intrafirm rates of diffusion of an innovation', *The Review of Economics and Statistics*, **45** (4), 348–59.

Mansfield, E. (1968), *Industrial Research and Technological Innovation: An Econometric Analysis*, New York: W.W. Norton.

Marchetti, C. (1979), 'Energy systems – the broader context', *Technological Forecasting and Social Change*, **14** (3), 191–203.

Marchetti, C. (1980), 'Society as a learning system: discovery, invention, and innovation cycles revised', *Technological Forecasting & Social Change*, **18** (4), 267–78.

Marchetti, C. (1988), 'Kondratiev's revisited – after one Kondratiev cycle', presentation for the International IIASA Conference on Regularities of Scientific–Technological Progress and Long-Term Tendencies of Economic Development, Novosibirsk, USSR, 14–19 March, pp. 1–8.

Marchetti, C. (2002), 'Productivity versus age', International Institute for Applied Systems Analysis report no. 00–155, Laxenburg, Austria.

Marchetti, C., P. S. Meyer and J. H. Ausubel (1995), 'Population dynamics: how much can be modeled and predicted', presentation for Program for Human Environment, Rockfeller University.

Marchetti, C., P. S. Meyer and J. H. Ausubel (1996), 'Human population dynamics revised with the logistic model', *Technological Forecasting and Social Change*, **52** (1), 1–30.

May, R. (1974), 'Biological populations with non overlapping generation stable points, stable cycles, and chaos', *Science*, **186**, 645–7.

McNeil, I. (1990), *An Encyclopedia of the History of Technology*, London: Routledge.

Metcalfe, J. S. (1970), 'The diffusion of innovations in the Lancashire textile industry', *Manchester School of Economics and Social Studies*, **2**, pp. 145–62.

Metcalfe, J. S. (1981), 'Impulse and diffusion in the study of technical change', *Futures*, **13** (5), 347–59.

Metcalfe, J. S. (1994), 'Industrial life cycles and the theory of retardation: towards an ecology of industrial development', *The Journal of Economics and Business Administration,* Kobe University, **170** (6), 41–61.

Ministry of International Trade and Industry, Japan (MITI) (2000), *Yearbook of Machinery Statistics, Chemical Industries Statistics and Steel Industries Statistics*, Tokyo: Research and Statistics Department, MITI (the older data are referred to in the back numbers).

Modis, T. M. (1992), *Predictions: Society's Telltale Signature Revealing the Past and the Future*, New York: Simon & Schuster.

Nakicenovic, N. and A. Grübler (1991), *Diffusion of Technologies and Social Behavior*, Berlin: Springer.

Nelson, R. and S. Winter (1982), *An Evolutionary Theory of Economic Change*, Cambridge, MA: Belknap.

Nihon Keizai Shinbunsha (1996), *The Silicon Valley Revolution*, Tokyo: Nihon Keizai Shinbunsha (Japan Economics Newspapers Inc.).

Rosenberg, N. (2001), *Schumpeter and Endogeneity of Technology: The Graz Schumpeter Lectures*, London: Routledge.

Schumpeter, J. A. (1939), *Business Cycles*, New York: McGraw-Hill.

Shimura, Y. (1992), *Semiconductor Industry in 2000*, Tokyo: Japan Management Association.

Stoneman, P. (1983), 'Theoretical approaches to the analysis of the diffusion of new technologies', in S. MacDonald, D. McL. Lamberton and T. Mandeville, *The Trouble with Technology: Explorations in the Process of Technological Change*, London: Pinter, pp. 93–103.

Takano, E. (1997), *Mechatronics*, Tokyo: Rikogaku Sha Inc.

United Nations (1996), *Industrial Statistics Yearbook* (the older data are referred to in the back numbers).

Utterback, J. M. (1994), *Mastering the Dynamics of Innovation: How Companies can Seize Opportunities in the Face of Technological Change*, Boston, MA: Harvard Business Press.

Vernon, R. (1966), 'International investment and international trade in the product cycle', *Quarterly Journal of Economics*, **80** (2), 190–207.

Walker, J. M. and E. B. Gingold (1985), *Molecular biology and biotechnology*, London: The Royal Society of Chemistry.

Yuasa, M. (1989), *Concise Science Chronological Table*, Tokyo: Sanseido.

Index

absorptive capacities 14, 63, 226–7
Achilladelis, B. 273
Adams, M. R. 187, 214
Advances in Social and Economic
        Aspects of Technology
        (ASEAT) 1
aerospace industry, services and 56–7
Alcamo, J. 254
Alchian, A. A. 31
Allansdottir, A. 154, 197
Almeida, P. 157
American Airlines 59
Amin, A. 156
Ancori, B. 81
Andersen, B. 273
Anderson, P. 181
Antonelli, C. 12, 19, 21, 36, 40, 41, 42
Anumba, C. J. 17
architecture, computer aided design
        (CAD) and 16–17
Arco 83, 84, 85, 87
Argyres, N. S. 35
Argyris, C. 14
Arora, A. 42, 153, 154, 156, 182
Arrow, K. J. 32, 41, 42, 226
Arthur, W. B. 254
Arundel, A. 156, 228
Arup 20, 23
AstraZeneca 57
Audretsch, D. B. 155
automobile industry
    crash testing 17–18
    fuel cell vehicle (FCV) 4–5, 126–48
        analysis 142–7
        fuel preference 126, 127, 133–42
        modelling R&D decisions 128–33
        opinion leaders and technological
                deviations 145–7
        rise of fuel cell vehicles 133–4
        technology choice at industry level
                142–5
    services and 56

banking
    distance banking 4
        Nordbanken case study 99–103,
                108–9, 116
        Société Générale case study
                109–11
    ICT in 92–3
    Internet banking 4, 92–122
        customers and development of
                117–18
        first mover advantages and
                disadvantages 94–6
        as innovation 96–9
        Nordbanken case study 4, 93,
                99–109, 116, 119–20
        Société Générale case study 4, 93,
                109–15, 116, 119–20
        suppliers and 118–19
Barbanti, P. 152, 153
Barker, T. 254
Barney, J. 13, 223
barriers to adoption 63
basic research 224–6
    bibliometric analysis 231–2
Bathelt, H. 157
Bauen, A. 127
Bayma, T. 156
Becker, G. S. 53
Beise, M. 223
Berg, R. D. 214
Bianchi, M. 52
Bilbao Guggenheim Museum 17
Bilda, Z. 22
biotechnology industry 152–76, 282–3,
        284, 286
    access to knowledge 162–6
    clustering and reaching out 154–8
        access to knowledge at distance
                156–7
        nature of distant search 157–8
        relative importance of local vs.
                distant relationships 154–6

distant networking strategies 158–9,
    171–4
establishing distant relationships
    168–70
food industry and 5–6, 179–218
    decomposing problem processing
        of innovations 183–6
    development of biotechnology in
        R&D 186–8
    discontinuities and distributed
        innovation 181–3
    emerging fusion of food and
        pharma R&D 208–10
    evolution of R&D themes
        through 1990s 194–6
    global distribution of LAB
        biotech innovations 197–8
    map of R&D themes 188–94,
        216–18
    methodology of study 186
    patent search procedures 215–16
    R&D profile of main actors in
        LAB biotechnology 202–5
    revealed roles in distributed
        innovation 198–202
    timing of patents 206–8
    types of actors and receptiveness
        to biotech opportunities
        196–210
    market relationships 166–8
    methodology of research on 159–61
    network structure 152, 153–4
BMW 143, 144
Bonaccorsi, A. 185
Bonazzi, G. 40
Boutellier, R. 198, 204
Bovee, M. 134
Boyer, R. 255
BP 77, 83, 84, 85, 87
Brennan, W. 154
Breschi, S. 155, 156
Bresnahan, T. F. 182
Brint, S. 131
Britton, S. 56
Brown, J. S. 14, 21, 81
Bryson, J. R. 56
Bucciarelli, L. 22
Buderi, R. 224, 225
Burgelman, R. A. 181
Burt, D. N. 63

Calantone, R. J. 65
capabilities approach 63–5, 74–5,
    78–83, 88
car industry *see* automobile industry
Chakrabarti, A. K. 231
Chandler, A. 2, 39, 40, 79, 80
Cheetham, P. S. J. 187, 203
Chesborough, H. 179, 181
Clark, K. B. 181
Clarysse, B. 166
cliometrics 254
Coase, Ronald 29, 30, 45
Cockburn, I. 179, 185, 199, 213, 228
codification 12
coercive pressures 129
cognition 22
Cohen, W. 14, 63, 223, 227, 228
Cohendet, P. 81, 156
collaboration 21
competition 254
    knowledge and 13
computer aided design (CAD) 16–17
computer based simulation 17
consumers (users)
    holistic view of consumption 64, 65
    innovation and 3–4, 51, 56–65, 63,
        97
        development of firm capabilities
            63–5
        Internet banking 117–18
    product development and 20
    service consumption 51, 52–6
    temporal dimension of consumption
        54–5
    utility and 54
Cooke, P. 152, 154, 155
Coombs, R. 179, 182, 212, 225, 227
Cosgel, M. M. 52, 53, 64
Coviello, N. E. 155
Cowan, R. 81
Cox, L. A. 260
craft skills 12
    data mining 16
crash testing 17–18
Crick, D. 155
Criqui, P. 259
Cyert, R. 13

Dachs, B. 217
Dahl, D. 22

DaimlerBenz/Chrysler, fuel cell vehicle
    (FCV) and 127, 133, 134, 137,
    138, 139–40, 143, 144, 145–6
Daniell, E. 187
Daniels, P. W. 56
databases 16
    data mining 16
David, P. 224, 254, 268
Davies, S. 268
Day, R. H. 254
de Vos, W. M. 215
decisionmaking 63
    expert systems and 18
deepwater exploration and production
    in petroleum industry 4, 73,
    75–8
    capabilities approach and 74–5,
        78–83, 88
    case studies 83–7
    path dependency 73, 85, 87
Delcour, J. 214
Demirkan, H. 22
Demsetz, H. 31
Denrell, J. 184
design 21
    complexity of 22
    computer aided design (CAD)
        16–17
    computer based simulation 17
    knowledge and 11
Dewick, P. 253
Dickson, K. 54
diffusion of innovation 6–7, 63, 267–96
    identification of innovation
        paradigm 280–3
    correlation of technology strategy
        and diffusion strategy 280–1
    structure of innovation paradigm
        281–3
    universities and 284–5
logistic dynamism 268–80
    bibliometric analysis of
        technology development period
        272–4
    description of technology
        development period 272
    determination of time span for
        innovation technologies 278–80
    determination of trajectory
        276–8

evidence of logistic nature of
    technology development 274–6
implications of 283–90
logistic equation 268–9
national systems of innovation
    288–90
product diffusion 269–72
technological opportunities and
    timing of venture business
    285–7
technology fusion by existing
    industries 290
technology trajectory and
    development trajectory 280
new insights 290–4
DiMaggio, P. J. 128, 129, 130
disposal services 58–9
distributed innovation 179
    discontinuities and 181–3
    global distribution of LAB biotech
        innovations 197–8
    revealed roles in 198–202
Dodgson, M. 11, 16, 19
Dosi, G. 2, 128, 130, 254
Dugas, B. 214
Duguid, P. 14, 21, 81
Duyk, G. 187
dyes, synthetic 273, 274, 283, 284, 285,
    286
dynamic capabilities theory 14

Earl, P. E. 53
Echeverri-Carroll, E. 154
Eckert, C. 22
Ehrenberg, E. 182
electronics industry 276–8, 281–2,
    284–5, 286, 289–90
Elf Aquitaine 77
Eliassaon, G. 184
enabling innovations 97
end-of-life disposal issues 59
e-science 16
evolutionary economics 13, 14, 84–5,
    254
expert systems 18
explicit knowledge 21
exploratory integration 173–4

Feldman, M. 156
Felsenstein, D. 154

Fiat 56
firms
  capabilities approach 74–5, 78–83
    63–65, 88
  commercialization of corporate
    science 6, 223–45
    basic research 224–6, 231–2
    conclusions 241–5
    cooperation, networking and
      knowledge flows 227–8
    diverging R&D output trends
      235–9
    information sources and
      methodology of study 231–5
    knowledge bases and absorptive
      capacity 226–7
    R&D output trends by industrial
      sector 239–41
    research papers in open literature
      229–31
    results of analyses 235–41
  consumption and innovation and
    3–4, 63–5
  innovation and consumption in
    51–65
  localized knowledge and 39–45
  modelling R&D decisions 128–33
    conceptual model 132–3
    firm-specificity and institutional
      entrepreneurship 131–2
    institutional change through the
      organizational field 130–1
    institutional embeddedness of
      organizations 128–30
  networking by *see* networking
  research by *see* research and
    development (R&D)
  strategic management 13
  technological opportunities and
    timing of venture business
    285–7
  theories of 13–14
    governance system 3, 33–9
    resource-based 3, 29, 32–3, 223
    transaction cost economics 29,
      30–2
first mover advantages and
    disadvantages, Internet banking
    94–6
Fisher, J. C. 268, 269

Fontes, M. 153, 158, 159, 160
food industry biotechnology 5–6,
    179–218
  decomposing problem processing of
    innovations 183–6
  development of biotechnology in
    R&D 186–8
  discontinuities and distributed
    innovation 181–3
  emerging fusion of food and pharma
    R&D 208–10
  lactic acid bacteria (LAB) 181, 186,
    187–8, 213–15
    evolution of R&D themes
      through 1990s 194–6
    global distribution of innovations
      197–8
    map of R&D themes 188–94,
      216–18
    R&D profile of main actors 202–5
    revealed roles in distributed
      innovation 198–202
    timing of patents 206–8
    types of actors and receptiveness
      to biotech opportunities
      196–210
  methodology of study 186
  patent search procedures 215–16
Foray, D. 224
Ford Motor Co 56, 145
Fornell, C. 95
Foss, N. J. 29, 32
Foxall, G. R. 63, 64
Freeman, C. 2, 180, 254, 255–8, 264,
    288
fuel cell vehicle (FCV) 4–5, 126–48
  fuel preference 126, 127, 133–42
    analysis 142–7
    fuel options and their
      consequences 134–6
    shifting preferences 136–42
  modelling R&D decisions 128–33
  opinion leaders and technological
    deviations 145–7
  rise of fuel cell vehicles 133–4
  technology choices at industry level
    142–5

Gadrey, J. F. 54
Gallaud, D. 156

Gambardella, A. 153, 156, 183
Gann, D. M. 17, 21, 23
Gardiner, P. 20
Garrelfs, R. 228
Gehry, Frank 17, 24
General Electric (GE) 56–7, 59
General Motors (GM), fuel cell vehicle
    (FCV) and 127, 138, 139, 140,
    143, 146–7
Georghiou, L. 225, 227
Gershuny, J. I. 51
Geullec, D. 236
Geuna, A. 12, 19, 21, 156
Gibbons, M. 20, 228
Giddens, A. 131
Godec, M. 77
Godin, B. 231
goods
    performance of 61–2
    service qualities of 55–6
    services encapsulated with 56–63
governance 3, 29–46
    role of localized knowledge 39–45
    system of 33–9
        general considerations 33–6
        model 36–9
    towards an economics of governance
        3, 30–3
        from resource-based theory to
            economics of governance 32–3,
            34
        from transaction costs economics
            to economics of governance
            30–1, 34
government research institutes (GRIs)
    205, 210
Granstrand, O. 212
Grant, R. M. 13, 223
Greenfield, H. I. 54
Greenlee, E. 231
Greenwood, R. 130
Grid, The 16
Griliches, Z. 223, 225, 268
Grübler, A. 259, 268
Grupp, H. 180
Gualerzi, D. 52
Guilhon, B. 42

Hagedoorn, J. 228
Hagiwara, T. 268, 270

Hall, B. 223
Halperin, M. R. 231
Hansen (Christian) Group 201, 202,
    203, 204
Harianto, F. 93, 95
Hart, D. 127
Haveman, H. A. 129
Hayes, D. 83
Hayward, A. B. 75, 76
Henderson, R. M. 181, 228
Hicks, D. 231
Hill, P. 54
Hinings, C. R. 130
Hirooka, M. 268, 269, 279, 290
Hoffman, A. J. 128, 130, 144
Höhlein, B. 127
Holt, K. 63
Howcroft, B. 93
Howell,, J. 56
Hsu, J.-Y. 154
Hutchins, E. 16

IBM, CATIA system 17
industrial revolutions, theory of 254–5
information and communications
        technology (ICT) 3
    banking and 92–3
    codification and 12
    innovation and 12, 14–15, 24–5
        assessment of impact and
            relevance 18–23
        knowledge management and
            15–18
innovation 1
    consumption (users) and 3–4, 51,
        56–65, 63, 97
        development of firm capabilities
            63–5
        Internet banking 117–18
    diffusion 6–7, 63, 267–96
        identification of innovation
            paradigm 280–3
        logistic dynamism 268–80,
            283–90
        new insights 290–4
    distributed 179
        discontinuities and 181–3
        global distribution of LAB
            biotech innovations 197–8
        revealed roles in 198–202

food industry
  decomposing problem processing
    of innovations 183–6
  discontinuities and distributed
    innovation 181–3
  global distribution of LAB
    biotech innovations 197–8
  revealed roles in distributed
    innovation 198–202
information and communications
  technology (ICT) and 12,
  14–15, 24–5
  assessment of impact and
    relevance 18–23
  knowledge management and
    15–18
  intensification 11
  Internet banking as 96–9
  national systems of 288–90
  paradigm 280–3
    universities and 284–5
  problem decomposability 183–6
  suppliers and 118–19
  theory of industrial revolutions
    254–5
institutional theory, modelling R&D
  decisions and 128–33
  conceptual model 132–3
  firm-specificity and institutional
    entrepreneurship 131–2
  institutional change through the
    organizational field 130–1
  institutional embeddedness of
    organizations 128–30
Internet 16
  banking 4, 92–122
  customers and development of
    117–18
  first mover advantages and
    disadvantages 94–6
  as innovation 96–9
  Nordbanken case study 4, 93,
    99–109, 116, 119–20
  Société Générale case study 4, 93,
    109–15, 116, 119–20
  suppliers and 118–19
Ismail, A. 15

Jaffe, A. 229
Jeffcoat, R. 187, 203

Jensen, R. L. 198, 204, 207
Jones, M. 155
Judson, H. F. 187

Kaku, S. 273, 274
Kalhammer, F. R. 133, 136, 138
Karabel, J. 131
Katz, J. S. 231
Kisaka, S. 277
knowledge 1
  absorptive capacities 14, 63, 226–7
  access to knowledge in
    biotechnology industry 156–7,
    162–6
  appropriation 224–31
  bases 226–7
  collaboration and 21
  creation of 20, 226–7
  design problems and 11
  dynamic capabilities theory 14
  explicit 21
  firms and 13
  flows 227–8
  learning theories and 14
  localized 39–45
  nonlinear dynamism of knowledge
    transfer 267–96
  as resource 13
  resource-based theory and 3, 32–3
  search for 157
  specialization and 13
  spillovers 44, 224–31
  tacit 21, 156, 158
  transactions costs economics and 31,
    42
Kondratiev, Nikolai 254, 279
Kondratiev waves 254, 258, 259, 264,
    268, 279, 280
Konings, W. N. 213
Kuipers, O. P. 215

Lamberton, D. 32
Lancaster, K. J. 53
Langlois, R. N. 52, 53, 64, 75, 79, 80,
    81, 82, 88
Larson, C. 225
Lazonick, W. 2
learning, theories of 14
Lemarié, S. 155, 174
Leonard-Barton, D. 79

Levin, R. C. 227
Levinthal, D. 14, 223, 227, 228
Levinthal, R. 63
Levy, D. L. 128
Li, T. 65
Lieberman, M. B. 95
Liebeskind, J. P. 179
Lim, K. 239
Linux 15
Lissoni, F. 155, 156
Loasby, B. J. 32, 36, 39, 52, 74, 78
localized knowledge 39–45
lock-in 254
long-run technical change 6, 253–65
    Freeman and Louçã's theory 255–8
    Kondratiev waves 254, 258, 259, 264,
        268, 279, 280
    simulation model 259–61
        capital accumulation 261
        demand and price determination
            261
        GDP growth 260
        investment function 261
        production function 259–60
        results 262
        supply function 260
    theory of industrial revolutions
        254–5
Louçã, F. 254, 255–8, 264
Lundvall, B. 63
Lynskey, M. J. 179

McCullogh, M. 12
McGown, A. 22
McKelvey, M. 99, 152, 154, 155
McNeil, I. 278
Mahnke, V. 32
Mahoney, Joseph T. 81
Mangematin, V. 155
Mansfield, E. 223, 225, 268
March, J. 13
Marchetti, C. 268, 269, 273, 274, 276
Margolis, J. 187
Marshall, Alfred 55
Martin, B. R. 227, 231
Maruo, K. 138
Mathé, H. 58
May, R. 294
Mazda 140
mediated integration 172–3

Metcalfe, J. S. 52, 268
Metcalfe, S. 179, 182, 212
Meuter, M. 97
Meyer-Krahmer, F. 81, 185, 228
Miles, I. D. 51
Millennium Footbridge (London) 23,
    24
Miller, F. 17
mimetic pressures 129
mining, virtual reality and 18
Miyazaki, K. 215
Modis, T. M. 268, 273
Moffat, L. A. R. 63
Montgomery, D. B. 95
Morange, M. 187, 198
Moritis, G. 83
Mott MacDonald 20
Mowery, D. C. 182, 226
Munro, H. J. 155

Nakicenovic, N. 268
national systems of innovation
    288–90
Nelson, R. R. 1, 13, 14, 128, 130, 131,
    226, 229, 293, 294
Nestlé 201, 202, 204, 209, 210
networking
    biotechnology industry 152, 153–4
        clustering and reaching out
            154–8
        distant networking strategies
            158–9, 171–4
        establishing distant relationships
            168–70
        market relationships 166–8
    distributed innovation 179
    knowledge flows and 227–8
Newman, R. 22
Nissan 138, 140
Noble, David 14–15
Noll, M. 217
Nonaka, I. 21
Nooteboom, B. 36, 40
Nordbanken 4, 93, 99–109
    distance banking 99–103, 108–9, 116
    Internet banking 99–100, 103–8,
        109, 116, 119–20
Nordhaus, W. 254
Normann, R. 96, 97, 99, 117
normative pressures 129–30

O'Farrell, P. N. 63
Oliver, C. 129, 130, 144
Organisation for Economic Co-
    operation and Development
    (OECD) 15
organizations *see* firms
Orsenigo, L. 153, 154
Owen Smith, J. 154
Oxman, R. 22

Pádua, M. 158
Pae, J. . 63, 65
Parkinson, S. T. 63
path dependency, deepwater
    exploration and production in
    petroleum industry 73, 85, 87
Pavitt, K. 12, 19, 96, 211–12, 226
Pearce, H. 139
Pennings, J. M. 93, 95
Penrose, Edith 1, 13, 29, 32, 64, 74, 79, 81
Perez, C. 2, 256, 259
pesticides 274–5
Petit, P. 54
Petrobras 76, 77, 85–6
petroleum industry, deepwater
    exploration and production 4,
    73, 75–8
    capabilities approach and 74–5,
    78–83, 88
    case studies 83–7
    path dependency 73, 85, 87
Petroski, H. 23
Pfeffer, J. 128
Pfirrmann, O. 166
pharmaceuticals industry
    fusion of food and pharma R&D
    208–10
    virtual reality and 18
Pisano, G. 13
Popken, D. A. 260
Porter, S. 22
Powell. W. 128, 129, 130, 152, 179
Prevezer, M. 155
Pridmore, R. D. 204
problem decomposability 183–6
Proctor & Gamble 20
productivity 64
Pry, R. H. 268, 269
purchasing 63
Pyka, A. 153

quality control 117–18
Queré, M. 153
Quinn, J. J. 54

Rees, K. 154
Reger, G. 20
Reid, G. 214
relationships *see* networking
relieving innovations 97
Renault 140
repair and maintenance services 58
research and development (R&D)
    absorptive capacities and 14
    basic research 224–6
    bibliometric analysis 231–2
    distributed innovation and 182–3
    food industry
        development of biotechnology in
        R&D 186–8
        emerging fusion of food and
        pharma R&D 208–10
        map of R&D themes 188–94,
        216–18
        R&D profile of main actors in
        LAB biotechnology 202–5
    modelling R&D decisions 128–33
        conceptual model 132–3
        firm-specificity and institutional
        entrepreneurship 131–2
        institutional change through the
        organizational field 130–1
        institutional embeddedness of
        organizations 128–30
    output trends 241–5
        diverging 235–9
        industrial sector 239–41
    research papers 229–31
resources
    knowledge as 13
    resource-based theories 3, 13, 29,
    223
        from resource-based theory to
        economics of governance 32–3,
        34
Rhodes, A. 84
Richardson, George 1, 64, 74, 75, 78,
    80, 81, 82, 88
Rip, A. 198
Roberfroid, M. B. 214
Roberts, J. 157

Robertson, J. D. 86
Robertson, P. L. 62, 64, 80
Robinson, J. 54
Robinson, W. T. 95
Rolls-Royce 57, 59
Romeo, J. B. 155
Romer, A. 23
Rosa, J. A. 99
Rose, B. 77
Roseboom, J. 198
Rosenberg, N. 224
Rosenkopf, L. 157
Roth, M. S. 155
Rothenberg, S. 128
Rothwell, P. 20
routines, evolutionary theory and 14
Ruffles, P. C. 58
Rutten, H. 198

Saarela, M. 214, 215
Salancik, G. 128
Salter, A. 21, 23, 197, 227
Sampson, G. P. 54
Sanchez, Ron 81
Saviotti, P. P. 153
Saxenian, A. 154
Schamalensee, R. 95
Schmoch, U. 185, 228
Schon, D. 14
Schrage, M. 11, 21, 23, 25
Schumpeter, J. A. 2, 254, 268, 279, 287
science, commercialization of 6, 223–45
    basic research 224–6
        bibliometric analysis 231–2
    cooperation, networking and knowledge flows 227–8
    information sources and methodology of study 231–5
        bibliometric analysis of basic research 231–2
        databases and definitions 232–4
        industrial sectors 234–5
    knowledge bases and absorptive capacity 226–7
    research papers in open literature 229–31
    results of analyses 235–41
        conclusions 241–5

        diverging R&D output trends 235–9
        output trends by industrial sector 239–41
Scitovsky, T. 62
Scott, W. R. 128, 129, 130, 131
Sekisui 20
self service technologies (SSTs) 97
Senge, P. M. 14
Senker, J. 179
services
    consumption and 51, 52–6
        innovation and 51, 56–65
    goods encapsulated with 56–63
    goods with service qualities 55–6
    Internet banking as service innovation 96–9
Shapiro, C. 94, 121
Shapiro, R. D. 58
Sharp, M. 179
Shaw, B. 63
Shell 76, 77, 137
Shirley, K. 76
Shostack, G. L. 56
Simon, H. 13, 31, 182, 183
simulation, computer based 17–18
Sjöberg, N. 182
Small, H. 231
Smith, K. 182
Snape, R. H. 54
social organization 22
Société Générale 4, 93
    distance banking 109–11, 116
    Internet banking 93, 109–15, 116
Soete, L. 180, 254
Sorenson, O. 155
Soukup, W. R. 63
specialization, knowledge and 13
SPICE system 18
spillovers 44, 224–31
Spulber, D. 41
Stacey, M. 22
Stahl, H. 223
Statoil 83, 84, 85, 87
Steinmuller, E. 12, 18, 21, 22, 156
Stephan, P. E. 155
Stigler, G. J. 53
Stoneman, P. 268
strategic management 13
Stuart, T. 155

suppliers, innovation and 118–19
Swann, G. M. Peter 55, 155

tacit knowledge 21, 156, 158
Takeuchi, H. 21
Tannock, G. W. 214
technical support services 58
Teece, D. J. 1, 13, 14, 36, 39, 42, 212, 227
Thomas, M. 83
Thomas, R. 63
Thomke, S. 11, 15, 17, 99
Tijssen, R. J. W. 223, 231, 239
Toffel, M. W. 59
Torre, A. 156
Tovey, M. 22, 23
Toyota 139, 145
training 58
Trajtenberg, M. 182
transactions costs economics 14, 29
    knowledge and 31, 42
    from transaction costs economics to economics of governance 30–1, 34
Tuomi, I. 15
turnkey operations 58
Tushman, M. L. 181

uncertainty 129
Unilever 201, 202, 204
universities 205, 237–9, 284–5
    The Grid 16
users *see* consumers (users)

utility 54
Utterback, J. M. 287

Valentin, F. 198, 203, 204, 207
Van den Hoed, R. 126, 134
van der Meulen, B. 198
Varian, H. R. 94, 121
Varma, R. 224
Vergragt, P. J. 126
Vernon,R. 290
Vincenti, W. G. 11
virtual reality 18
von Hippel, E. 20, 62, 63, 99

Wagner, D. R. 214
Walsh, J. 156
Watts, P. 76, 77, 78
web services 15
Whyte, J. K. 17, 18, 22, 25
Williams, R. 20, 24
Williamson, Oliver E. 14, 29, 30, 31, 45
Winter, S. G. 1, 13, 14, 128, 130, 293, 294
Wood, P. A. 56
Wright, M. 93
Wynstra, F. 63

Yielding, C. A. 87
Yu, T. F. 62, 65
Yuasa, M. 278

Zeller, C. 155
Zucker, L. 155, 156, 230